Mechanics and calculations
of
textile machinery

Mechanics and calculations of textile machinery

N. Gokarneshan
B. Varadarajan
C. B. Senthil Kumar

WOODHEAD PUBLISHING INDIA PVT LTD
New Delhi ● Cambridge ● Oxford ● Philadelphia

Published by Woodhead Publishing India Pvt. Ltd.
Woodhead Publishing India Pvt. Ltd., G-2, Vardaan House, 7/28, Ansari Road
Daryaganj, New Delhi – 110002, India
www.woodheadpublishingindia.com

Woodhead Publishing Limited, 80 High Street, Sawston, Cambridge,
CB22 3HJ UK

Woodhead Publishing USA 1518 Walnut Street, Suite1100, Philadelphia

www.woodheadpublishing.com

First published 2013, Woodhead Publishing India Pvt. Ltd.
© Woodhead Publishing India Pvt. Ltd., 2013
Reprint, 2020

Woodhead Publishing India Pvt. Ltd. ISBN: 978-93-80308-20-3
Woodhead Publishing Ltd. ISBN: 978-0-85709-104-8
Woodhead Publishing Ltd. e-ISBN: 978-0-85709-552-7

Typeset by Sunshine Graphics, New Delhi
Printed and bound in India by Replika Press Pvt. Ltd.

Dedicated to
Our beloved parents
Our revered gurus
and
Our beloved students

Contents

PART II
Calculations of textile machinery

Preface

It gives me immense pleasure to bring out the first edition of the book which is much required for students of diploma and undergraduate courses in textile technology. The contents of the book have been carefully structured so as to meet the curriculum requirements of undergraduate course. A number of worked out examples appear at various parts of the text. One part of the book deals with mechanics and the other deals with fundamental calculations. The text matter is also supported by many illustrations which are beneficial to the readers. Even though a number of books are available relating to the subject, no book, however, is able to satisfactorily meet the curriculum requirements of the course. The book thus has clear advantage over the existing books in this aspect. Moreover, lucid explanations are provided at appropriate places, enabling the book to be more reader friendly. Even though the book has been prepared to the best of my knowledge and my co-authors have put in their best efforts, there still exists a good scope for improvement in the content. In this regard, I take it an opportunity to invite valuable suggestions from readers for enhancing the quality of the book. Error if any is regretted. I also take it an opportunity to thank our beloved chairman and mentor, Sri Raja M Shanmugam, our vice-chairmen, Sri CMN Muruganandam, and Sri P Moghan, and chairman (academic development committee), Sri S Dhananjeyan, for their encouragement and moral support.

Dr N Gokarneshan

PART I

Mechanics of textile machinery

1

Belts and rope drives

1.1　Introduction

Power source is always required to process the textile materials in order to convert it into a final product. An electric power is converted to rotational energy by means of electric motor and it is transferred to the machines to perform various jobs. Power can be transmitted between two parts or shafts, either by negative or positive means. In the case of negative method of power transmission there is slippage, whereas in the positive one there is no slippage. The belts and rope drives are negative methods of power transmission. This chapter deals in detail about the various methods of belt and rope drives and highlights their merits and demerits.

1.2　Various methods of drive

Generally machines are driven by the following two methods:

1. Individual drive
2. Group drive

1.2.1　Individual drive

The motor may drive the machine shaft directly through a coupling, rope, belt, chain and gears. Hydraulic actuators, D.C. motors, A.C. motors or Stepping motors are used for machine drives.

1.2.2　Group drive

A prime mover, either a diesel engine or electric motor drives a common shaft running through a shed from which all the machines gets its drive by a belt or rope.

1.3   Applications of belt and rope drives in textile industry

(a)   Drives in blow room: motor to mono-cylinder, ERM beater, mixing-bale-opener and stripping and take-off rollers in unimixer.
(b)   Drives from top coiler to base coiler plate of modern carding machine.
(c)   Main drives in all spinning, texturing machines and compressors.
(d)   Motor to licker-in and cylinder.
(e)   Cleaner roller of stripper roller at the delivery side.
(f)   Motor to flat-stripper roller.
(g)   Crossed-flat-belt drive from cylinder to a pulley from where further drive (consisting of clutch, worm/worm gear) to the driving-shaft of flat in a modern card.
(h)   Drive to drafting rollers and other rolling elements in a single delivery draw frame. The use of flat belts instead of timing belts or gears reduces periodic faults. The fibre, dirt accumulation on the teeth of toothed wheels used in timing belt drive increases the periodic faults.
(i)   Drives to opening rollers, friction drums, and take-off rollers of DREF spinning machine.
(j)   Drive to rotor in a rotor-spinning machine.
(k)   Main drive from motor in draw-texturing machine.
(l)   Drive to creel-rollers of a modern draw frame.
(m) Tape drives used in driving spindles in a group at ring spinning, ring doubling and TFO.
(n)   Flat belt drives used in driving spindles in a group at ring spinning and TFO.
(o)   Variator pulley drives at ring spinning and more.

1.4   Transmissions of power by belts and ropes

The belts and ropes transmit power through the friction between belt or rope and rim of the pulley. The function of a belt and rope drive is to transmit rotational motion and torque from one shaft to another, smoothly, quietly and inexpensively. Belt and rope drives provide the best overall combination of design flexibility, low cost, low maintenance, ease of assembly and space savings.

Compared to other forms of power transmission, belt drives have the following advantages:

(a)   They are less expensive than gear or chain drives.
(b)   They have flexible shaft centre distances, where gear drives are restricted.
(c)   They operate smoothly and with less noise at high speeds.

(d)   They can be designed to slip when an overload occurs in the machine.
(e)   They require no lubrication, as do chains and gears.
(f)   They can be used in more than one plane.
(g)   They are easy to assemble and install and have flexible tolerances.
(h)   They require little maintenance.
(i)   They do well in absorbing shock loading.

They do not have an indefinite life. While in use, they need regular inspection schedule to guard against wear, aging and loss of elasticity.

There is some slip and creep in the belts, and so the angular velocity ratio between the driver and driven is neither constant nor equal to the ratio of the pulley diameters (exemption – timing belts). During working, the belt tension is sag on one side of the driving pulley and tight on another side. The belt should be arranged such that the tight side is on the bottom side and slack side on top of the pulleys. This is shown in the Fig. 1.1. Otherwise, the angle of contact between the belt and rim of the pulley reduces, decreasing the power transmission of the belt.

1.5   Different methods to adjust the belt tensions

When belts are in operation, they stretch over time. Machines that utilize a belt drive need some feature that can compensate for the belt stretch, such as an adjustable motor base, or an idler pulley.

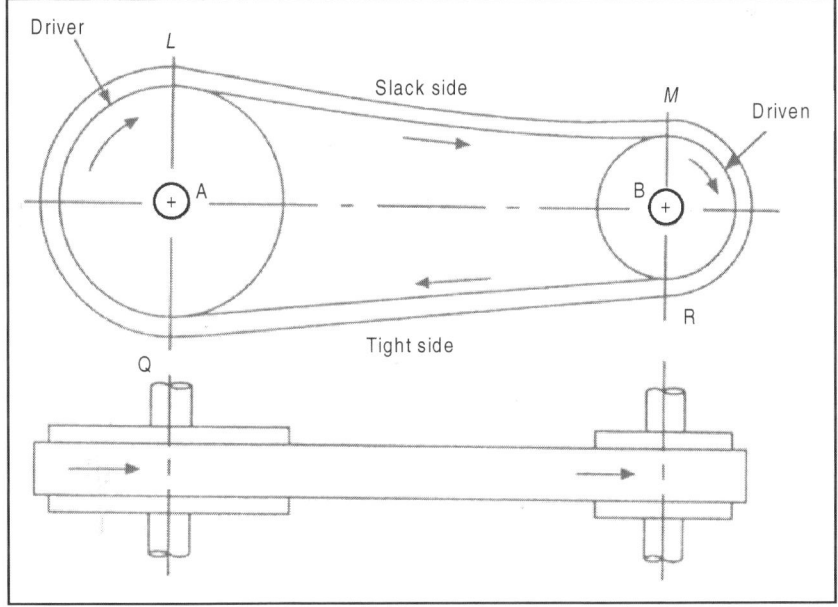

1.1 Slack and tight side of belt.

An *idler pulley* is used to maintain constant tension on the belt. It is usually placed on the slack side of the belt and is preloaded, usually with springs, to keep the belt tight (Fig. 1.2).

In the case of flat belts with joints or hinges, a short length of belt is cut periodically to remove the slackness in the belts.

For endless belts (some flat belts and V belts), the centre distance between pulleys is slightly increased by means of an adjusting screw. In case of drive from motor to main shaft of a machine, provision is made to move the motor away from the driven pulley through adjusting screws on the bed of the motor as shown in Fig. 1.3.

Other way of adjusting the belt tension is through mounting the motor on to a bed with pivot as shown in Fig. 1.4.

1.6   Types of belts

Belts are typically made of continuous construction of materials, such as rubberized fabric, rubberized cord, reinforced plastic, leather, and fabric (i.e., cotton or synthetic fabric).

There are six types of belts as listed below:

1. Flat belt	4. Cog belt
2. V-belt	5. Timing belt
3. Multi-V-belt	6. Round belt or rope drive

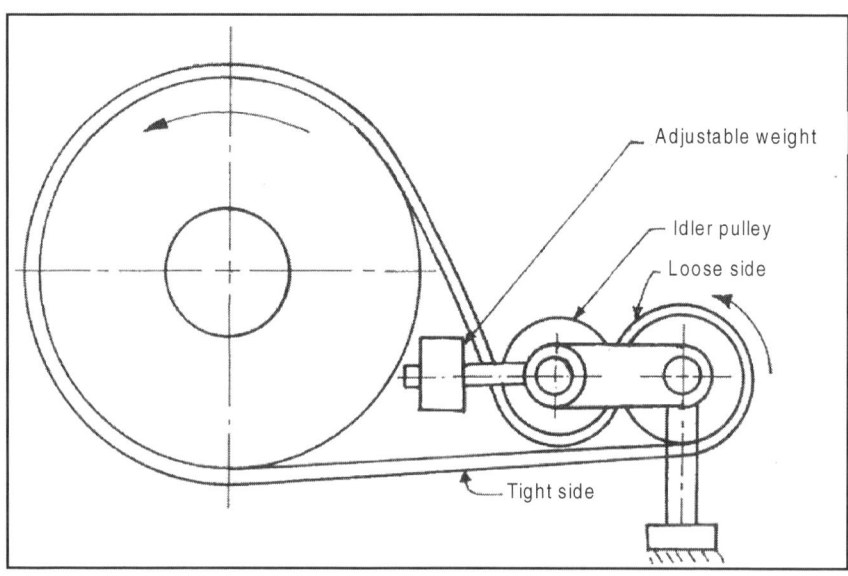

1.2 Idler pulley with adjustment weight.

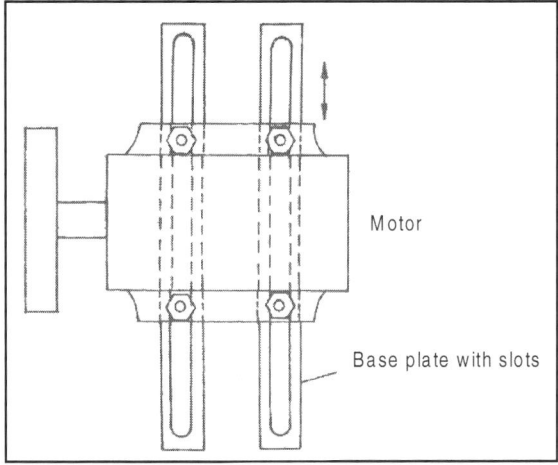

1.3 Adjusting screws on motor bed.

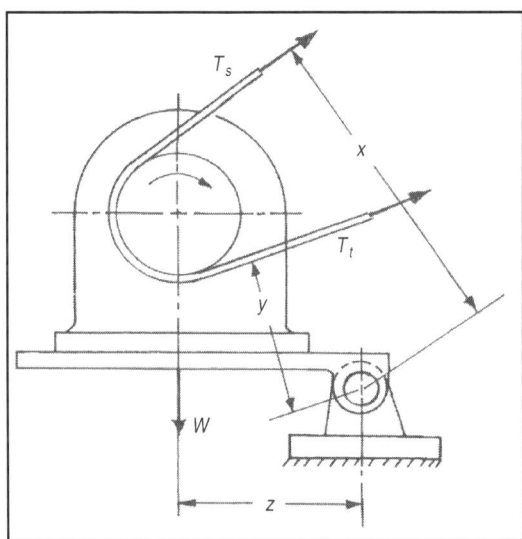

1.4 Motor bed with pivot.

Types of belt-drive

There are two common types of belt drives: (a) *open-belt drive,* and (b) *crossed-belt drive.* In the open-belt drive the driver and the follower move in the same direction. While in the crossed-belt drive, the sense of rotation of the driven pulley must be opposite to that of the driving pulley. These two arrangements, illustrated in Fig. 1.6, are used to connect shafts which are parallel.

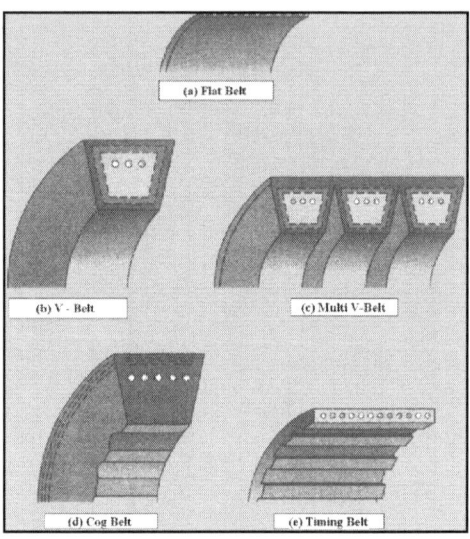

1.5 Types of belts.

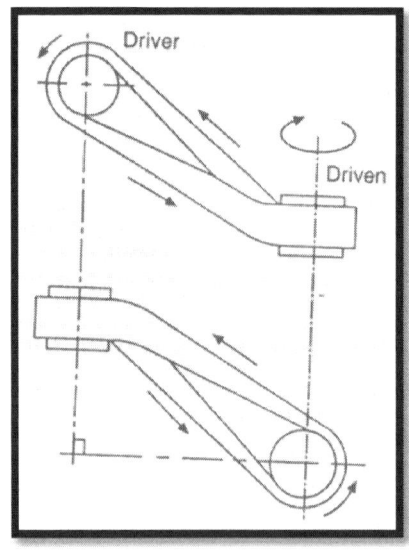

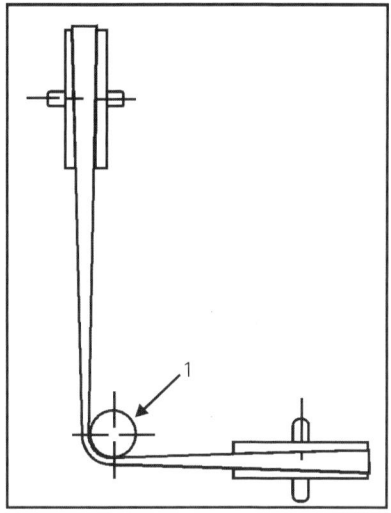

1.6 (a) Quarter-turn drive.   (b) Right angle drive.

1.6.1   Flat belt

A *flat belt* is shown in Fig. 1.5(a), Fig. 1.7 and Fig. 1.8. It is the simplest type but is typically limited to low-torque applications because the driving force is restricted to pure friction between the belt and the pulley. The

1.7 Flat pulley and belt.

1.8 Flat belt application in textile mills: Carding machine; Wool and cotton cards.

properties such as flexibility, durability, strength of the belt and high coefficient of friction between the belt and the rim of the pulleys are to be considered for the construction of flat belts.

Group-drive systems were in use in olden days, which drives several machines through pulleys and flat belts. Earlier, flat leather belts were available in two varieties, viz., oak-tanned and mineral or chrome-tanned. They were made by using few layers of leather cemented together to get the required thickness. Commercial leather belts were specified according to the number of layers, such as single-, two-, three- and four-ply belts. These belts offer moderate coefficient of friction between rim of the pulley and the belt. Over a period of time, they become rigid and exhibit creep. Also they have poor resistance against moisture.

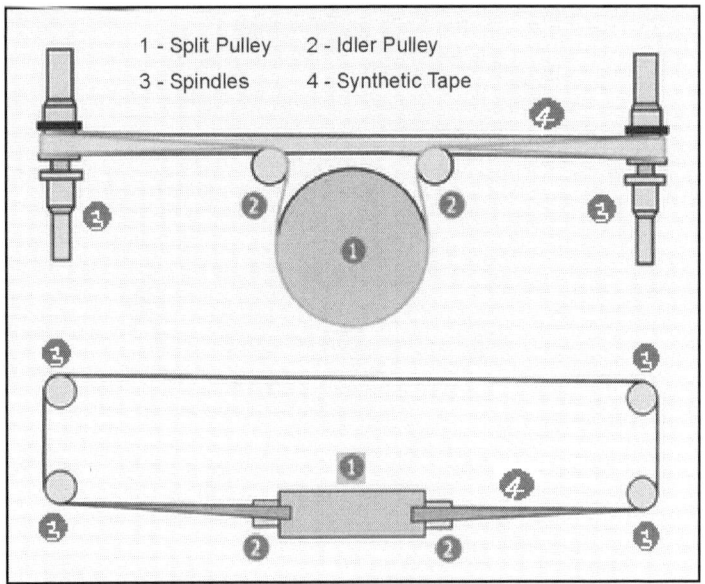

1 - Split Pulley   2 - Idler Pulley
3 - Spindles   4 - Synthetic Tape

1.9 Flat tape drive in spinning/doubling/TFO machines.

Nowadays, flat belts are made of urethane or rubber matrix, reinforced with fabric or nylon cords or steel wire. They are strong and durable to withstand higher speeds. Flat belts are quite, efficient at high speeds, and can transmit large amounts of power over long distances. The coefficients of friction of leather, polyamide and urethane flat belt are 0.4, 0.5 to 0.8 and 0.7, respectively.

Geometry of belt drive

Open and crossed-flat belt drives

The length of the belt in an open belt drive is found by the following equation:

$$L = 2C + [\pi (D_1 + D_S)/2] + [(D_1 - D_S)^2/4C] \qquad \dots (1)$$

For crossed belt as shown in Fig. 1.12, the following equation is valid.

$$L = 2C + [\pi (D_1 + D_S)/2] + [(D_1 - D_S)^2/4C] \qquad \dots (2)$$

$$\text{As } \acute{\alpha} = \acute{\alpha}_S = \acute{\alpha}_1 = 2[180 - \cos^{-1}\{(D_1 + D_S)/2C\}] \qquad \dots (3)$$

Power ratings

Power rating of flat belt depends upon its maximum possible tension. This

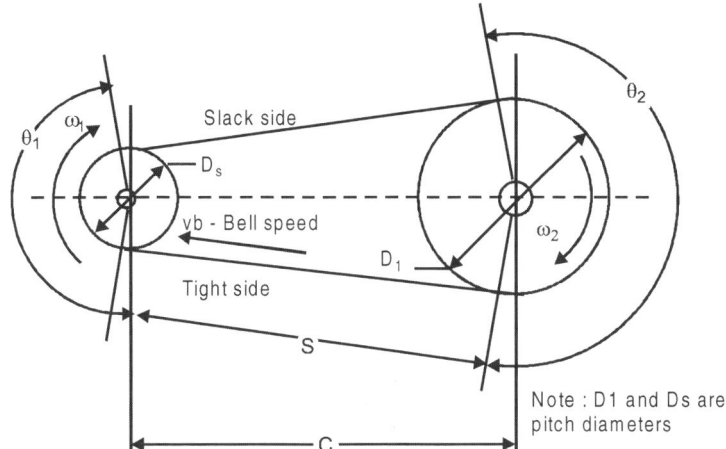

1.10 Geometry of belt drive.

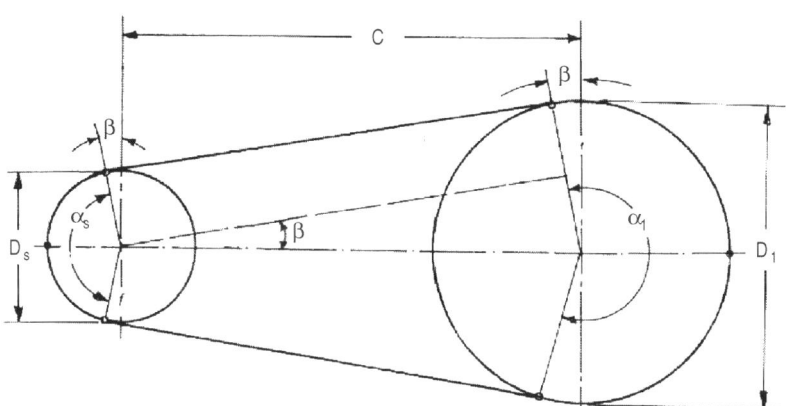

1.11 Open belt drive.

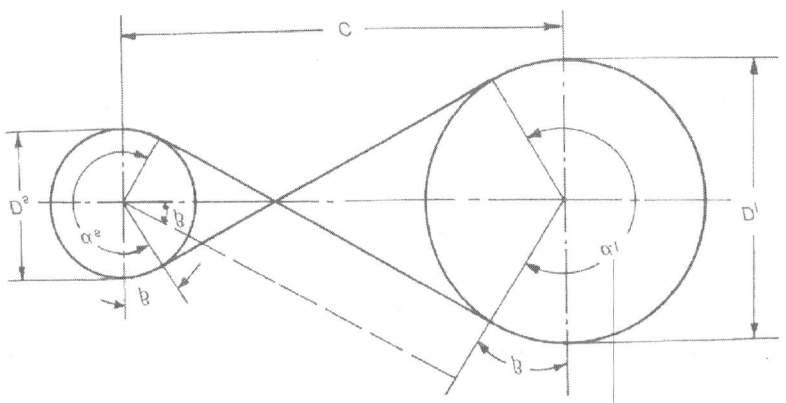

1.12 Cross belt drive.

is obtained from the equation:

$$P_T = (T_A \times v \times F_P \times F_V) \div K$$

where, P_T = Power transmitted; T_A = 2Ti = Allowable belt tension (N); Ti = Initial belt tension (N); v = Belt speed (m/sec); F_P = Pulley or belt correction factor; F_V = Velocity correction factor; K = Belt service factor.

Maximum power transmission of belt

(a) Power transmission by belts

$$\text{Power in kW} = (T_1 - T_2)\, y/1000$$

where, T_1 and T_2 = Belt tensions in N, y = Belt speed in m/s.

$$= 2\pi\, NM_t / 60{,}000$$

where, M_t = Torque in Nm and $M_t = (T_1 - T_s) \times D/2$; D = Diameter of pulley in metres.

(b) Relation between T_1 and T_s

$$T_1/T_s = e^{\mu\theta},$$

Where, μ – co-efficient of friction and θ – angle of overlap over smaller pulley in radians. [Radian = $\theta\pi/180°$]

(c) Centrifugal tension in a belt

$T_c = \rho v^2$, π = mass of belt per unit length in kg/m.
$T_c = \rho v^2 A$, ρ = mass of belt per unit volume in kg/m^3.
A = Area of cross-section of belt in m^2.

$$\text{Initial tension} = (T_1 + T_s) / 2 \text{ [When } T_c \text{ is negligible]}$$

$$= (T_1 + T_s + 2T_c) / 2 \text{ [When } T_c \text{ is considered]}$$

$$T_{max} = T_1 + T_s$$

(d) For absolute maximum power

$$T_c = T_{max}/3$$

From the above equation, optimum belt speed for absolute maximum power can be calculated.

(e) Rope or V-belt drive

$T_1/T_s = e^{[\mu\theta/\sin\alpha]}$, where, α = semi-groove angle.

Selection of pulley diameter

The diameter of the driving pulley can be determined from the expression

$$D = 1.114 \times \sqrt[3]{(P \div N)} \, mm$$

where, D = diameter of the driving pulley; P = power transmitted in kW; N = rpm of the driving pulley.

A pulley can be selected from the belt manufacturer's catalogue. A pulley whose size is nearest to the dimension as per the design calculation may be selected. As a result, the belt velocity and the size of the driven pulley would also vary.

Special purpose belt drives

Flat belts with cone drums or cone pulleys are used for variable speed drives in blow rooms and speed frames as shown in Fig. 1.25. The belt is moved axially for varying the output speed. For stepped pulleys, V-belt or round belt can be used with grooved sheaves as shown in Fig. 1.26. The stepped pulleys with V-belts are commonly used in main drives of many textile machines.

Speed adjustment of driven pulley is possible with a variator-drive using adjustable grooved/conical discs and V-belt without varying the speed of the motor. A variator drive used in ring frame is shown in Fig. 1.27. By shifting the driver and driven discs axially, the effective diameters of the discs over which the axis of the belt passes are simultaneously varied, thus varying the output speed. To increase the output speed or spindle speed, the driver discs are moved closer to each other and the driven discs moved apart and vice versa. A control device through hydraulic or pneumatic pistons and lever mechanisms effect the movement of the discs. The spindle speed can be varied in several steps depending on the doff-position and the end-breakage rate of the yarn. This permits higher throughput of yarn as optimum spindle speed is selected at any instant.

1.6.2   V-belt

A V-belt is shown in Fig. 1.5(b). It is the most widely used type of belt, particularly in automotive and industrial machines. V-belts comprise cord tensile members located at the pitch line, embedded in a relatively soft matrix which is encased in a wear-resistant cover. V-belts are generally made of rubber covered with rubber or polymer impregnated fabric and reinforced with nylon or polyester, glass, aramid or steel wires/cords. The wedging action of a V-belt in a pulley groove results in a drive which is more compact than a flat belt drive, but short centre V-belt drives are not conducive to shock absorption. The V shape causes the belt to wedge tightly into the pulley, increasing friction and allowing higher operating torque.

A V-belt or rope meshing in a V-shaped groove of the pulley increases greatly the frictional resistance to slipping, for a given maximum belt tension, as compared to flat belt. The rope or V-belt does not rest on the bottom of the groove but wedges itself into the groove.

The cross-section of a V-belt is trapezoidal as shown in Fig. 1.13. The thickness b ranges from 8 to 19 mm and the width ranges from 12.5 to 38 mm. The belt section angle may be around 40°. The V-belts are available in A, B, C, D, E sections. The width, thickness, minimum sheave diameter increase from section A to E, in other words, belts become heavier from section A to E.

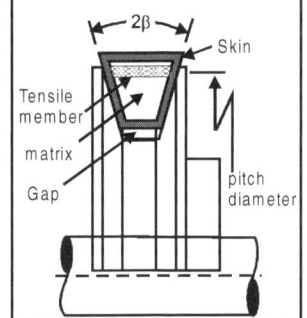

1.13 V- belt and grooved pulley.

1.6.3   Multi-V-belt

A multi-V-belt is shown in Fig. 1.5(c) and Fig. 1.14. Its design is identical to several V-belts placed side by side but is integrally connected. It is used to increase the amount of power transferred. If a single V-belt is inadequate for power transmission, then multiple belts and corresponding multi-grooved pulleys (Fig. 1.15) are necessary; this pulley is equipped with a tapered bush for axle clamping without the stress concentration associated with a key.

1.14 Multi-V-belt.

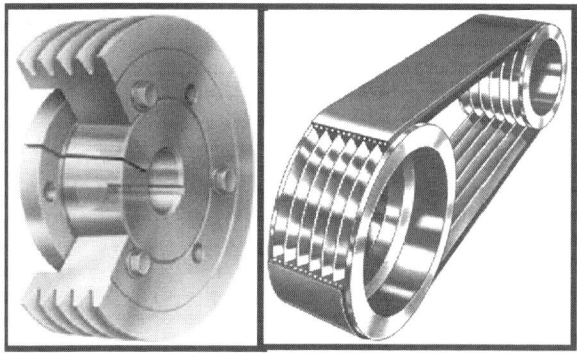

1.15 Multi-grooved pulleys.

The rather extreme short-centre drive above illustrates a problem with multiple belts – how to ensure equitable load sharing between flexible belts whose as-manufactured dimensional tolerances are significantly looser than those of machined components, for example.

1.6.4   Cog belt

A cog belt is shown in Fig. 1.5 and Fig. 1.16. Its design is similar to a V-belt but has grooves formed on the inner surface known as cogging which alleviate deleterious bending stresses as the belt is forced to conform to pulley curvature. This feature increases belt flexibility, allowing the belt to turn smaller radii, and hence can be used on smaller pulleys, reducing the size of the drive. The belt illustrated also incorporates slots on the underside.

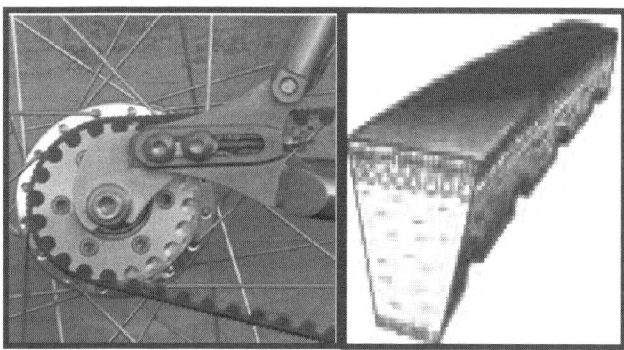

1.16 Cog belt and pulley.

1.6.5   Timing belt

A *timing belt* is shown in Fig. 1.17. Its design has teeth-like gear that engages with mating teeth on the pulleys. It combines the flexibility of a belt with the positive grip of a gear drive. It is widely used in applications where relative positioning of the respective shafts is desired.

1.17 Timing belt.

These are positive drives using flat-type belts. The belts have flat outer surface and evenly spaced teeth on the inner surface and operate on toothed pulleys. The toothed pulley looks like a spur gear as shown in Fig. 1.18. Timing belts offer very good accuracy in transmitting motion compared to flat belts and are comparable to gears. In addition they offer greater flexibility in the location of driver and driven.

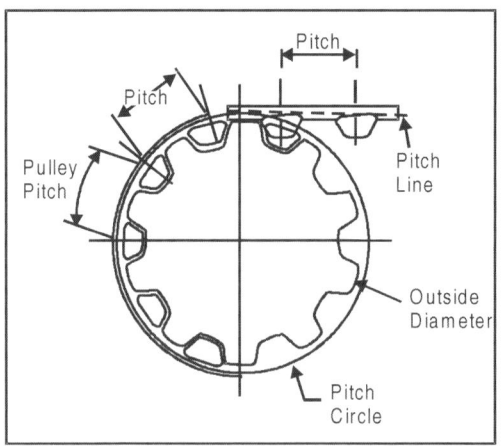

1.18 Timing-belt drive.

They are commonly used in high-speed machines, when the distance between driver and driven is of considerable length. In these situations they offer greater advantage over gear drives in terms of lower power consumption and noise. Usage of gear drive system under these circumstances involves placement of several carrier wheels/gears (gear trains), which leads to high power consumption due to their own weights and also leads to noise. In addition, the gear drive system becomes so complicated that changing of certain gears (change wheels) to effect a change in the process parameters or for maintenance purpose is difficult.

Examples of use of timing belts in textile machinery are texturing machines and rolling elements at the delivery side of modem carding machine, such as doffer, stripper, calendar and coiler rollers. Drive between top coiler plates of a combing machine is through timing belts. Required belt tensions in timing belts are low and, consequently, bearing loads are reduced. Timing belts find application in drives for changing the direction of yarn twist in ring frame.

1.19 Timing belt drive in an air-jet texturing machine.

A timing belt drive used in an air-jet texturing machine is shown in Fig. 1.19. The belt is guided over driving and driven wheels by two discs (guards) placed on each side of these wheels. This is to avoid the belt slipping out of the wheels due to misalignment of the wheels. The belt is kept under tension by means of a tension wheel, which does not have side discs.

Timing belts find application in drives for changing the direction of yarn twist in ring frame. Timing belt is made of rubberized fabric reinforced with steel wires to take on the tension. The steel wire is located at the pitch line and the pitch length is the same regardless of the

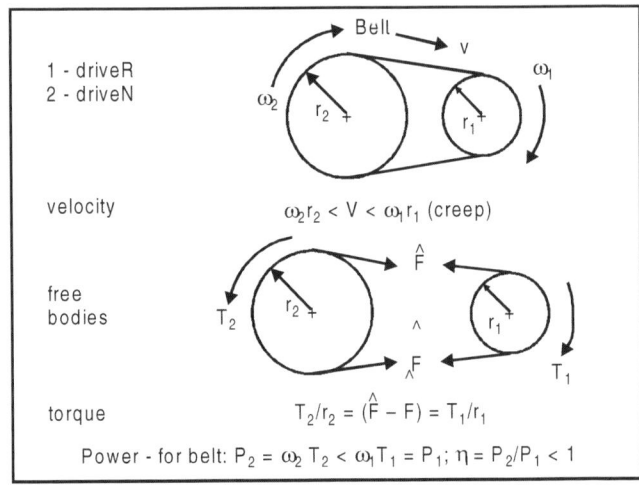

1.20 Kinematics and kinetics of a belt drive.

The kinematics and kinetics of a belt drive

Figure 1.20 shows the kinematics and kinetics of a belt wrapped around two pulleys. The transfer of power in a belt drive relies critically on friction. The tensions F_{min} and F_{max} in the two *strands* (the nominally straight parts of the belt not in contact with the pulleys) cause a normal pressure over the belt–pulley contact, and it is the corresponding distributed friction whose moment about the pulley centre equilibrates the shaft torque T – provided gross slip of the belt on the pulley surface does not occur due to friction breakaway. The speed reduction ratio and the torque amplification ratio are each equal to the radius ratio, so that the output power equals the input power and the efficiency is 100%.

The torque ratio equals the ideal ratio (as may be seen from the free bodies), but *creep* results in the speed ratio being less than ideal. Creep – not to be confused with gross slip – is due to belt elements changing length as they travel between F_{min} and F_{max}, and since the pulley is rigid then there must be relative motion between belt element and pulley. Since power equals the product of torque and (angular) speed, the consequence of the foregoing is that efficiencies of real belts are less than 100%.

1.6.6   Rope drive belt

These belts are circular in cross-section, without any joints. They can be used for transmission over a long distance. The pulley on which it runs should have V-shaped groove (Fig. 1.21). For variable speed drives using stepped pulleys, this belt can also be used. When all pulleys (driving, driven

1.21 Rope drive belt.

and guide pulleys) are at considerable distances, and at different planes, these belts can be used.

A rope meshing in a V-shaped groove of the pulley increases greatly the frictional resistance to slipping, for a given maximum belt tension, as compared to flat belt. The rope does not rest on the bottom of the groove but wedges itself into the groove (Fig. 1.22).

1.7   Rope drive

This type of drive is used in blow room for driving evener roller and stripper roller and in carding for driving the flat cleaning brush rollers, burnishing rollers, metallic clothing grinding equipment, etc.

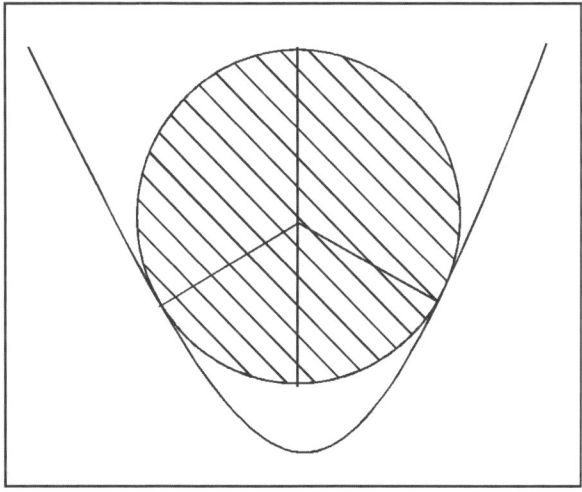

1.22 Wedging of rope inside pulley groove.

1.23 Rope drive.

Steel or *wire ropes* are used for transmission of power in cases where the parts to be connected are at a large distance apart, and where extra strength is needed (Fig. 1.23). They are used in lifts, colliery winding and hauling arrangements, mill drives, etc. The ropes run on grooved pulleys; but contrary to the practice adapted with cotton ropes, they rest on the bottom of the grooves and are not wedged between the sides of the grooves.

1.7.1   Advantages of rope drive

Following are the advantages of a rope drive over belt drive:

1. The rope drive is particularly suitable, when the distance between the shafts is large.
2. The frictional grip in case of rope drive is more than that in the belt drive.
3. The net driving tension (i.e., difference between the two tensions) in the case of rope drive is more than the belt drive (because the ratio of tensions in the case of rope drive is cosec α times more than that in the belt drives.)

1.7.2   Ratio of tensions in rope drive

Consider a rope running in a groove as shown in Fig. 1.24.

Let R_1 = Normal reactions between rope and sides of the groove R = Total reaction in the plane of the groove 2α = Angle of the groove μ = Coefficient of friction between rope and sides of the groove

Resolving the reactions vertically to the groove, $R = R_1 \sin \alpha + R_1 \sin \alpha = 2 R_1 \sin \alpha$ or $R_1 = (R/2 \sin \alpha)$

We know that the frictional force = $2\mu.R_1 = 2\mu \times (R/\sin \alpha) = \mu.R/\sin \alpha = \mu.R \operatorname{cosec} \alpha$

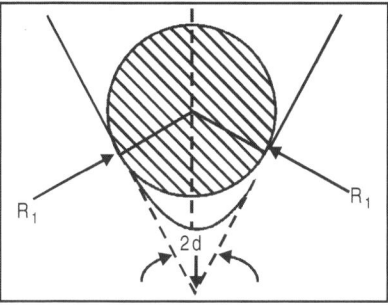

1.24 Rope running in a groove.

Now consider a small portion of the belt subtending an angle $\delta\theta$ at the centre. The tension on one side will be T and on the other side $T + \delta T$. Now we get the frictional resistance equal to $\mu.R$ cosec $\acute{\alpha}$ against the relation between T_1 and T_2 for the rove drive will be 2.3 log $(T_1/T_2) = \mu.\theta$ cosec $\acute{\alpha}$.

1.8   HP transmitted by ropes

HP transmitted by ropes $= (P\ v\ N) \div 4500$

$$= [(T_1 - T_2)\ v\ N] \div 4500$$

where P = driving force in kgf, v = the velocity in m per minute, and N = the number of ropes in the pulley.

The relation between two tensions T_1 and T_2 is given by

$$T_1 \div T_2 = e^{\ [\mu\theta\,/\,Sin\alpha]} \qquad\qquad \dots (a)$$

where 'e' is the base of the napierian log = 2.718, μ = coefficient of friction for the rope on the pulley, θ = angle of lap in radians and 2 α = angle of the groove.

Taking log on both sides, the equation (a) becomes log $T_1 \div T_2 = 0.4343$ $\times$ [$\mu\theta$ / sin α] in circular measure $= 0.007578 \times [\mu\theta$ / sinα] if θ is in degrees (b)

1.9   Comparison of flat and V-belts

The advantages of flat belts in comparison to V-belts are listed below:

(1) They are simple in design and are relatively inexpensive.
(2) They can be easily maintained in terms of periodic adjustment of belt tension and their replacement when worn out.
(3) Precise alignment of pulleys and shafts is not so critical.
(4) They give better protection to the machinery against impact or

overloads. As they are flexible and long, they have the ability to absorb shock and vibrations due to slipping action.

(5) Clutching action with flat belt is possible by moving it from fast to loose pulley and vice versa. This type of arrangement is commonly found in old blow rooms, cards, draw frames and speed frames.

(6) Using cone pulleys, different velocity can be obtained for the driven element by moving the flat belt axially.

(7) They can be used for long distances, even up to 15 m, where other types of drives cannot be used.

The major advantages of V-belt are as follows:

(1) V belts are used for short distance, which results in compact construction.

(2) Due to wedge action between the belt and the pulley, the slip is less.

(3) Wedging action permits a smaller arc of contact, increases the pulling capacity of the belt and consequently results in an increase in the power transmission capacity.

(4) They can be used for high-speed reduction up to 7:1.

(5) They can be operated even when the belt is vertical.

(6) They are made available in endless form, which results in smooth and quite operation, even at high speeds.

The major disadvantages of V-belts are as follows:

(1) The construction of V-grooved pulleys is complicated and costlier compared to the pulleys for flat belt drives.

(2) The creep in V-belts is higher compared to flat belts.

(3) The ratio of thickness of V-belts to pulley diameter is high, which increases the bending stress in the belt cross-section and adversely affects its durability.

1.10   Variable speed drives

For variable speed drives in blow rooms and speed frames *flat* belts with cone drums or cone-pulleys are used as shown in Fig. 1.25. For varying the output speed, the belt is moved axially. For stepped pulleys, V-belt or round belt can be used with grooved sheaves as shown in Fig. 1.26. The stepped pulleys with V-belts are commonly used in main drives of many textile machines.

Speed adjustment of driven pulley is possible with a variator drive using adjustable grooved / conical discs and V-belt without varying the speed of the motor. A variator drive used in ring frame is shown in Fig. 1.27. By shifting the driver and driven discs axially, the effective diameters of the

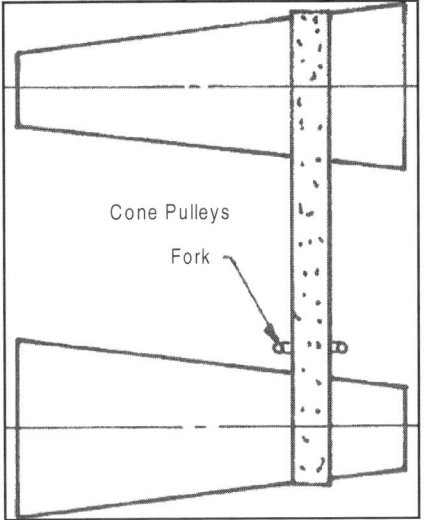

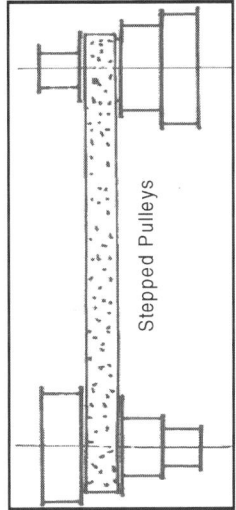

1.25 Cone pulley.   *1.26* Step pulley.

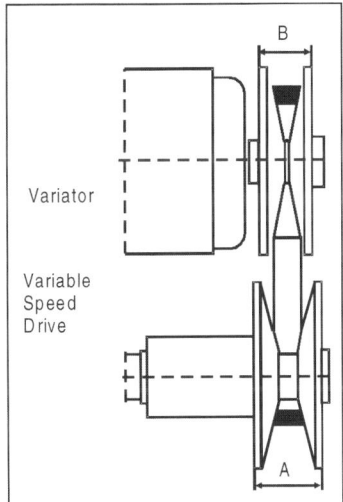

Fig. 1.27 Variator drive.

discs over which the axis of the belt passes are simultaneously varied, thus varying the output speed. To increase the output speed/spindle speed, the driver discs are moved closer to each other and the driven discs moved apart and vice versa. A control device through hydraulic or pneumatic pistons and lever mechanisms effect the movement of the discs. The spindle speed can be varied in several steps depending on the doff-position and the end-breakage rate of the yarn. This permits higher throughput of yarn as optimum spindle speed is selected at any instant.

Reeves variable speed transmission in which the operator is able to control the speed while running without any interruption of the work in hand, and the change is continuous instead of being abrupt. This consists of a pair of pulleys connected by a V-shaped belt in the manner indicated in Fig. 1.25. Each pulley consists of a pair of driving disks with cone-shaped faces, the disks revolving with the shafts, but at the same time capable of sliding longitudinally along it to a certain extent. To adjust the diameters of the pulleys, the two conical disks on one shafts forming a pair are caused to approach each other, virtually increasing the diameter or recede from each other when the diameter is reduced. It follows that when the disks of one pair are approaching each other, those of the opposite pair must automatically be made to recede to the same extent. In this manner the ratio of driving diameter to driven diameter is readily and quickly changed, thus securing any desired speed without the necessity of stopping the machine.

Variable speed transmission in the case of a chain drive may be secured by a device similar to that described above using PIV (positive, infinitely variable) gear which has radial teeth in the conical disk.

1.11   Centrifugal tension

We have already discussed that the belt continuously runs over both the pulleys. It carries some centrifugal force in the belt, at both the pulleys, whose effect is to increase the tension on both, tight as well as the slack sides. The tension caused by the centrifugal force is called centrifugal tension. At lower speeds, the centrifugal tension is very small and may be neglected. But at higher speeds, its effect is considerable, and thus should be taken into account.

Consider a small portion AB of the belt as shown in Fig. 1.28,

Let m = mass of the belt per unit length

v = linear velocity of the belt

r = radius of the pulley over which the belt runs

T_c = centrifugal tension acting tangentially at P and Q

$d\theta$ = angle subtended by the belt AB at the centre of the pulley

Therefore, length of the belt AB = $r.d\theta$ and mass M = $m.r.d\theta$
We know that centrifugal force of the belt AB,

$$P_c = Mv^2 / r = (m.r.d\theta) \, v^2 / r = m.d\theta.v^2$$

Now resolving the forces (i.e., centrifugal force and centrifugal tension) horizontally and equating the same,

$$2T_c \sin (d.\theta/2) = m.d\theta.v^2$$

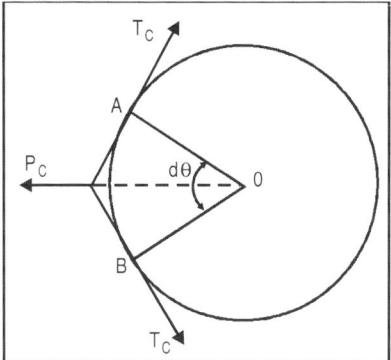

1.28 Rope or round belt.

Since $d\theta$ is very small, therefore substituting sin $(d.\theta/2) = d.\theta/2$ the above equation,

$$2T_c\,(d.\theta/2) = m.d\theta.v^2 \text{ or } T_c = m.v^2$$

Notes

1. When the centrifugal tension is taken into account, the total tension, on the tight side $= T_1 + T_c$ and total tension on the slack side is $T_2 + T_c$.
2. The centrifugal tension on the belt has no effect on the power transmitted by it. The reason for the same is that while calculating the power transmitted, we have to use the values = total tension on tight side − total tension on the slack side

$$(T_1 + T_c) - (T_2 + T_c) = (T_1 - T_2)$$

1.11.1 Maximum tension in the belt

Consider a belt transmitting power from the driver to the follower. Let f = maximum safe stress in the belt in N/mm^2, b = width of the belt in mm, and t = thickness of the belt in mm.

We know that maximum tension in the belt,

T = maximum stress × cross-sectional area of belt

When centrifugal tension is neglected, then maximum tension

$$T = T_1$$

and when centrifugal tension is considered, then maximum tension

$$T = T_1 + T_c$$

1.11.2 Condition for transmission of maximum power

The power transmitted by a belt,

$$P = (T_1 - T_2) \times v \text{ W} \qquad \ldots \text{(a)}$$

where T_1 = tension on the tight side in N, T_2 = tension on the slack side in N, and v = velocity of the belt in m/s

The ratio of tensions, $T_1 \div T_2 = e^{\mu\theta}$ or $T_2 = T_1 \div e^{\mu\theta}$

Substiting the value of T_2 in equation (a),

$$P = (T_1 - T_1 / e^{\mu\theta}) \times v = T_1 (1 - 1 / e^{\mu\theta}) \times v = T_1 \times v \times C \text{ (b)}$$

where $C = (1 - 1 / e^{\mu\theta})$

We know that tension on the tight side is $T_1 = T - T_c$

where T = maximum tension in the belt in Newton, and
T_c = centrifugal tension in Newton.

Substituting the value of T_1 in equation (b),

$$P = (T - T_c) v C = (T - mv^2) v C = (Tv - mv^2) C$$

We know that for maximum power, differentiating the above equation and equating the same to zero,

$$T - 3 mv^2 = 0 \text{ (c)}$$

$$T - 3T_c = 0 \text{ (substituting m.v}^2 = T) \text{ or } T = 3T_c$$

It shows that when the power transmitted is maximum, 1/3rd of the maximum tension is absorbed as centrifugal tension.

1.11.3 Belt speed for maximum power

For maximum power transmission,

$$T - 3mv^2 = 0 \text{ or } 3mv^2 = T$$

$$\therefore \qquad v = (T / 3m)^{1/2}$$

Where,
v = speed of the belt for maximum transmission of power,
T = *maximum tension in the belt,
m = mass of the belt for unit length.

Note: The power transmitted when 1/3 of the maximum tension is absorbed as centrifugal tension (condition of last article) at belt speed for

---

*   Maximum tension in the belt is equal to the sum of tensions on the tight side (T_1) and centrifugal tension (T_c).

maximum power (condition of the above article) is known as absolute maximum power or in other words, maximum power which can be transmitted under any conditions.

1.12   Selection of flat belt

Given data: (1) Power and speed in rpm of the driver unit. (2) Speed in rpm of the driven unit. (3) Space limitations and operating conditions.

Procedure for selection

Step 1   Calculate the speed ratio

Step 2   Assume the belt speed, V, within the range given.

Step 3   Determine the diameters of the pulleys using the belt speed and rpm of the respective pulleys

Step 4   Round off the pulley diameters to the recommended values given, satisfying the following conditions:

    (a)   Allowable variation in the speed of driven pulley = ±10%

    (b)   The actual belt speed, calculated using the speed and diameter of the driving pulley, must be within the range

Step 5:   Find out correction factor for arc of contact

Step 6:   Find out load correction factor

Step 7:   Calculate corrected load (or) Design power, using the following equations:

    Corrected load (or) Design power = Power to be transmitted × Load correction factor × Arc of contact factor

Step 8:   Select the type of belt

Step 9:   Write down the load rating of the selected belt, at the belt speed of 10 m/s

Step 10:   Calculate the load rating of the belt at the actual belt speed, V.

Step 11 :   Determine the required mm plies of the belt

Step 12:   Assume minimum no. of plies (= 3) and determine the required width of the belt

Step 13:   For the type of belt selected and the no. of plies assumed, select the required width of belt. If the required width is not standard, choose a wider belt.

    If a belt to suit the requirements cannot be selected for the no. of plies assumed, assume a higher number of plies and repeat steps 12 and 13

Step 14:   Calculate the belt length

Step 15:   Determine the length by which the belt must be shortened to provide correct initial tension

Problem 1.1: Select a flat belt from the manufacturer's catalogue to transmit 11 kW at 1200 rpm from an engine to a line shaft at 450 rpm. The maximum centre distance between the shafts is 2 m.

1. Speed ratio = 1200 ÷ 450 = 2.66
2. Belt speed = 20 m/s (assumed)
3. To find the diameter of pulleys,
   $20 = (\pi Dn) \div 60$;
   $D = (60 \ v) \div n\pi$
   $D_1 = (60 \times 20) \div 450 = 0.849 = 849$ mm.
   $D_2 = (60 \times 20) \div 1200\pi = 0.318 = 318$ mm
4. Standard diameters are 800 and 315 mm
   (a) Speed of driven pulley $N_2 = 1200 \times (315 \div 800) = 472.50$ rpm
       % variation in the speed of the driven pulley
       $$= \frac{472.50 - 450 \times 100}{450} = 5$$
   (b) Actual belt speed (calculated using the speed and the diameter of the driving pulley)
       $$V = \frac{\pi D_1 N_1}{60} = \frac{\pi \times 0.315 \times 1200}{60}$$ (belt speed is within the range)
       $V = 19.79$ m/s
5. Arc of contact
   $$= 180 - \frac{D - d}{C} \times 60$$
   $$= 180 - \frac{800 - 315}{2000} \times 60$$
   $= 180 - 14.55$
   $= 165.45°$
   Arc of contact factor = 1.0582
6. Load correction factor: 1.5 (Shock loads assumed)
7. Design power = 11 × 1.5 × 1.0582 = 17.46 kW
8. Type of belt selected = FORT
9. Load rating at 1 cm/s = 0.0289 kW/mm/ply
10. Load rating at 19.79 m/s = $0.0289 \times \dfrac{19.79}{10}$ = 0.0572 kW/mm/ply
11. Plies of belt (in mm) = $\dfrac{17.46}{0.0572}$ = 305.24

12. Belt width (in mm) = $\dfrac{305.24}{3}$ = 101.75 3

This width is not available in FORT for three ply. Therefore 4 ply is tried.

Belt width = $\dfrac{305.24}{4}$ = 76.06 mm

In 4 ply, 90 mm width belt is available in FORT.

13. So, 4 ply 90 mm belt width is selected.
14. Belt length (assuming open drive),

$$L = 2C + \frac{\pi(D+d)}{2} + \frac{(D-d)^2}{4C}$$

$$= 2 \times 2000 + \frac{\pi}{2}(800+315) + \frac{(800+315)^2}{4 \times 2000}$$

$$= 4000 + 1751.44 + 29.4$$

$$= 5780.84 \text{ mm}$$

15. For initial belt tension, 1% of the belt length should be reduced. Therefore length of the belt = 5780.84 − 5780.84 × 1 / 100 = 5723.03 mm FORT 4 ply, 90 mm belt width of 5723 mm length is selected.

1.13   Selection of V-belt

Selection procedure

Given data – Power to be transmitted, speed of the driver unit, speed of the driven unit, space limitations, operating conditions.

(A) Determination of pulley diameters, center distance and belt length

Step 1: Select standard V-belt section based on the kW rating.

Step 2: Determine the "Optimum belt speed range" within which the belt gives maximum power rating.

Step 3: Assume suitable belt speed less than 25 m/s within the "optimum belt speed range".

Step 4: Determine the diameters of the pulleys using the belt speed and rpm of the pulleys.

Step 5: Round off the pulley diameters to R20 series satisfying the following conditions.

    a) The diameter of the smaller pulley must be greater than the minimum value recommended.

    b) The % change in the rpm of the driven pulley must be within ±5.

c) The actual belt speed, if calculated using the diameter and rpm of the driving pulley, must be less than 25 m/s.

Step 6: Determine the C/D ratio corresponding to the actual speed ratio and calculate the center distance.

Step 7: Calculate minimum center distance (C min) and maximum center distance (C max.). Also check whether the center distance calculated in step 6 lies between minimum and maximum. If not, adjust the center distance suitably.

Step 8: Calculate belt pitch length. Standardize the belt pitch length.

Step 9: Calculate the center distance for the standard pitch length of the belt. Check whether it lies between C and if not, select new standard pitch length for the belt and repeat Step 9 till C lies between C min and C max.

(B) Determination of number of belts

Step 10: Select correction factor for industrial service, Fa.

Step 11: Find out small diameter factor, Fb.

Step 12: Find out belt length correction factor Fc. Corresponding to 'L' (Assuming value).

Step 13: Find out correction factor for arc of contact, Fd.

Step 14: Determine equivalent pitch diameter.

Step 15: Find out the power rating for the selected cross-section of the belt, at the actual belt speed, against 'd_e'.

Step 16: Determine the number of belts required to transmit the power. Round off the value to the next higher integer.

Step 17: Write down the specification of the belt selected.

Step 18: Determine the number of consecutive grading numbers any of which may be used to make a matched set.

Problem 1.2: Select V-belts to transmit 11 kW from an AC Motor, single phase, running at 1440 rpm to a piston compressor running at 900 rpm for more than 10 h per day. Space available for the drive is 2 m × 2 m.

Solution:

Step 1:   To transmit 11 kW, either B or C cross-section belt can be selected. C Cross-section belt is selected.

Step 2:   Optimum belt speed range within which the belt gives maximum power rating, for C cross-section belt is 17 m/s to 28 m/s.

Step 3:   The belt speed must be below 25 m/s. Otherwise, the pulley must be of special construction and further it will require dynamic balancing. So belt speed of 20 m/s is assumed.

Step 4: To determine diameter of pulleys.

$$V = \frac{\pi dn_1}{60} \; ; \; 20 = \frac{\pi d \times 1440}{60} \; ; \; d = 265 \text{ mm}$$

$$D = \frac{d \times n_1 \times \eta}{n_2} = \frac{265 \times 1440 \times 0.98}{900} = 439 \text{ mm}$$

Step 5: Standard pulley diameters: d = 280 mm and D = 450 mm (OR) d = 250 mm and D = 400 mm can be selected. For this problem, d = 280 mm and D = 450 mm is selected.

   a) The diameter of the smaller pulley is greater than the minimum value (200 mm) recommended.

   b) The percentage change in the rpm of the driven pulley

$$n_2 = d\frac{n_1}{D}\eta$$

$$= \frac{280}{450} \times 1440 \times 0.98 = 878.08 \text{ rpm}$$

   % change in rpm of the driven pulley = 900 – 878.08 × 100
   = 2.44 < ± 5

   c) The actual belt speed $V = \dfrac{\pi dn_1}{60 \times 1000}$

$$= \frac{\pi \times 280 \times 1440}{60 \times 1000} = 21.11 \text{m/s}$$

   Note: The actual belt speed is calculated using the diameter and rpm of the driving pulley.

   The actual belt speed is <25 m/s. The standard pulley diameters satisfy all the three conditions.

Step 6: (A) To determine C/D ratio:

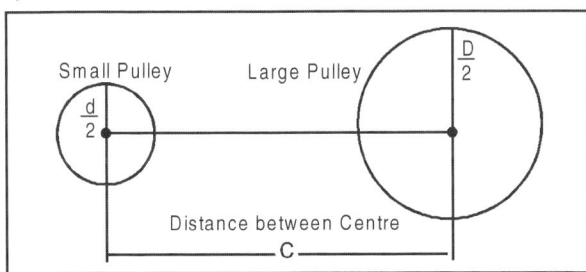

Actual speed ratio standard = diameter ratio

$$\frac{D}{d} = \frac{450}{280} = 1.607$$

Therefore, corresponding C/D ratio by interpolation

$$= \frac{1.5 - 0.3 \times 0.607}{1} = 1.5 - 0.1821 = 1.3179$$

C = 1.3179 × D = 1.3179 × 450 = 593.05 mm.

Step 7: C min = 0.55 (D + d) + T
= 0.55 (450+280) +14 = 415.5 mm
C max = 2 (D + d) = 2 (450 + 280) 1460 mm
The centre distance calculated in Step 6 lies between C min and C max. The two pulleys must be accommodated within the 2 m space available.
The space occupied by the two pulleys = C + (D + d)/2 = 593.05 + 450 + 280 = 948.05 mm < 2000 mm

Step 8: Belt pitch length (assuming open belt drive)
L = 2C + π/2 (D + d) + (D – d)²/4C
= 2 × 593+ π/2 (450 + 280) + (450 – 280)²/(4 × 593)
= 2344.86 mm
Nearest standard belt pitch length per C section belt = 2342 mm.

Step 9: Centre distance for a given belt length and diameter of pulleys
C = A+ √ A² – B

$$A = (L/A) - \left[\pi(D+d)\right]/8 = \frac{2342}{4} - \frac{\pi 450 + 280}{8}$$

= 585.5 – 286.67 = 298.83 mm
B = (D – d)²/8 = (450 – 280)²/8 = 3612.5 mm
C = 298.83 +v (298.83)² – 3612.5 = 591.55 mm

(B) Determination of no. of belts

Step 10: For piston compressors, driven by an AC single phase motor, running for more than 10 h/day. Industrial service factor, F_a = 1.5

Step 11: Small diameter factor, F_b = 1.12, corresponding to D/d ratio of 1.607.

Step 12: Belt length correction factor, F_c = 0.90

Step 13: Arc of contact =180 – 60 [D – d]/C = 180 – 60 (450 – 280)/ 591.55 = 162.76°; F_d = 0.9592 (by interpolation)

Step 14: Equivalent pitch diameter, $d_c = d_p \times F_b$ = 280 × 1.12 = 313.6 mm, where d_p is pitch diameter of smaller pulley.

Step 15: Power rating of C cross-section belt at actual belt speed of 21.11 m/s and equivalent pitch diameter of 313.6 mm = 11.33 kW.

Step 16: No. of belts required to transmit power

$$= \frac{P \times F_a}{kW \times F_e \times F_d} = \frac{11 \times 1.5}{11.33 \times 0.91 \times 0.9592} = 1.6 \simeq 2.$$

So, two belts are required. [Where, F_a = Correction factor; F_e = Correction factor length; F_d = Correction factor Arc of Contact]

Step 17: Designation of the V-belt selected C 2286/90 IS:2494

Step 18: No. of consecutive grading numbers any of which may be used to make a matched set = 4.

Problem 1.3: A 10 kW motor running at 1750 rpm has a pulley of 160 mm diameter fitted to it. It drives a line-shaft at a speed of 800 rpm. Three machines are driving by the line-shaft, their speed being 300, 500 and 200 rpm. The driving pulleys of the machines are respectively 240, 320 and 400 mm in diameter. Find the size of the pulleys to be fitted on to the line-shaft.

Here D_1 = 160, n_1 = 1750, n_2 = 800, D_2 = ?

$D_2 = D_1 \times n_1 / n_2 = (160 \times 1750) \div 800 = 350$ mm.

D_2 is the diameter of the pulley on the line-shaft through which power from the motor is transmitted.

Let d_1, d_2, and d_3 be the diameters of the pulleys fitted on the line shaft for driving machines.

$d_1 \times 800 = 300 \times 240 = 90$ mm

$d_2 \times 800 = 500 \times 320 = 200$ mm

$d_3 \times 800 = 200 \times 400 = 100$ mm

Problem 1.4: An motor shaft running at 120 rpm is required to drive a machine shaft by means of belt. The pulley on the motor shaft is of 2 m diameter and that of the machine shaft is of 1 m diameter. If the belt thickness is 5 mm, find the speed of the machine shaft when (i) there is no slip, and (ii) there is a slip of 3%.

Solution: Given, N_1 = 120 rpm; d_1= 2 m; d_2 = 1 m; t = 5 mm = 0.005 m; s = 3%.

(i) *Speed of the machine shaft when there is no slip.* We know that speed of the machine shaft,

$$N_2 = N_1 \times \frac{d_1 + t}{d_2 + t} = 120 \times \frac{2 + 0.005}{1 + 0.005} = 239.4 \text{ rpm}$$

(ii) *Speed of the machine when there is a slip of 3%.* We know that speed of the machine shaft,

$$N_2 = N_1 \times \frac{d_1 + t}{d_2 + t} \times \frac{(1 - 3)}{100} = 120 \times \frac{2 + 0.005}{1 + 0.005} \times \frac{(1 - 3)}{100} = 232 \text{ rpm}$$

Problem 1.5: The width of a belt is 150 mm and the maximum tension per mm of width is not to exceed 1.6 kg. The ratio of tension on the two sides is 2¼, the diameter of the driver 1 m, and it makes 220 rpm. Find the horse power that can be transmitted.

Solution: In this case, T_1 = 1.6 × 150 = 240 kg and T_1/T_2 = 2.25 or T_2 = 240 ÷ 2.25 = 106.7 kg.

Therefore, P = $T_1 - T_2$ = (240 – 106.7) = 133.3 kg.

We know that V = π d n = 3.142 × 1 × 220 m/min.

Therefore, hp = $\dfrac{133.3 \times 3.142 \times 220}{4500}$ = 20.5

Ratio of driving tensions in a belt: The ratio of driving tensions in a belt just on the point of slipping is given by

$$\frac{T_1}{T_2} = e^{\mu\theta}$$

where **e** is the base of the log = 2.718, μ = coefficient of friction for the belt on the pulley, and θ = angle of lap or arc of circumference embraced by the belt in radians.

Taking log on both sides, the equation becomes

$$\log T_1/T_2 = \mu\ \theta \log e$$

$$= 0.434\ \pi\ \theta \text{ in circular measure}$$
$$= 0.007578\ \mu\ \theta \text{ if } \theta \text{ is in degrees } 1.6$$

Problem 1.5: Find the length of belt necessary to drive a pulley of 500 mm diameter running parallel at a distance of 12 m from the driving pulley of diameter 1600 mm.

Solution: Given, d_2 = 500 mm = 0.5 m or r_2 = 0.25 m; 1 = 12 m; d_1 = 1600 mm = 1.6 m or r_1 = 0.8 m.

In this example, no mention has been made whether the belt is open or crossed. Therefore we shall find out the value of length of the belt in both the cases.

Length of the belt if it is open

We know the length of the belt if it is open,

$$L = \pi (r_1 + r_2) + 2l + (r_1 - r_2)^2/ l = \pi (0.8 + 0.25) + 2 \times 12 +$$
$$(0.8 - 0.25)^2 / 12 = 27.32 \text{ m}$$

Length of the belt if it is crossed

We know that length of the belt if it is crossed,

$$L = \pi (r_1 + r_2) + 2l + (r_1 + r_2)^2/ l = \pi (0.8 + 0.25) + 2 \times 12 +$$
$$(0.8 + 0.25)^2 / 12 = 27.39 \text{ m}$$

Problem 1.6: Find the length of the belt required for driving two pulleys in a cross belt drive of 600 mm and 300 mm diameter when 3.5 m apart. Take thickness of the belt as 5 mm.

Solution: Given: d_1 = 600 + 5 = 605 mm = 0.605 m or r_1 = 0.3025 m; d_2 = 300 + 5 = 305 mm = 0.305 m or r_2 = 0.1525 m; l = 3.5 m
We know that length of the cross belt drive

$$L = \pi (r_1 + r_2) + 2l + (r_1 + r_2)^2/ l$$

$$= \pi (0.3025 + 0.1525) + 2 \times 3.5 + (0.3025 + 0.1525)^2 / 3.5$$

$$= 8.488 \text{ m}$$

Problem 1.7: The tensions in the two sides of the belt are 1000 and 800 N, respectively. If the speed of the belt is 75 m/s, find the power transmitted by the belt.

Solution: Given, T_1 =1000 N; T_2 = 800 N; v = 75 m/s
We know that power transmitted by the belt,

$$P = (T_1 - T_2) \times v$$

$$= (1000 - 800) \times 75$$

$$= 15,000 \text{ N–m/s} = 15000 \text{ W} = 15 \text{ kW}$$

Problem 1.8: Find the necessary difference in tensions in kgf in the two sides of a belt drive, when transmitting 20 hp at 30 m/s.

Solution: Given, P = 20 hp; v = 30 m/s

Let $(T_1 - T_2)$ = Necessary difference in tensions in the two sides of the belt.

Problem 1.9 A laminated belt 8 mm thick and 150 mm wide drives a pulley of 12 m diameter at 180 rpm. The angle of lap is 190° and mass of the belt material is 1000 kg/m². If the stress in the belt is not to exceed 1.5 N/mm² and the coefficient of friction between the belt and the pulley is 0.3, determine the power transmitted when the centrifugal tension is (i) considered, and (ii) neglected.

Solution: Given, t = 8 mm = 0.008 m; b = 150 mm = 0.15 m; d = 1.2 m; N = 180 rpm; θ = 190° = 190 π/180 = 3.316 rad; π = 1000 kg/m³; f = 1.5 N/mm²; μ = 0.3

(i) Power transmitted when the centrifugal tension is considered

Let T_1 = Tension in the tight side of the belt, and
     T_2 = Tension in the slack side of the belt.
We know that speed of the belt,
     v = π × 1.2 × 180/60 = 11.31 m/s
Maximum tension in the belt,
     T = fbt = 1.5 × 150 × 8 = 1800 N and
mass of the belt per metre length,
     m = area × length × density
         (0.008 × 0.15) × (1) × (1000) = 1.2 kg/m
Therefore, centrifugal tension,
     T_c = m.v² = 1.2 (11.31)² = 153.5 N
and tension in the tight side
     T_1 = T – T_c = 1800 – 153.5 = 1646.5 N
We also know that

$$2.3 \log (T_1 / T_2) = \mu.\theta = 0.3 \times 3.316$$
$$= 0.9948; \log (T_1 / T_2)$$
$$= 0.9948 \div 2.3 = 0.4325$$

Therefore, 1646.5 / T_2 = 2.707 (Taking antilog of 0•4325) or T_2 = 1646.5/ 2.707 = 608.2 N and power transmitted,

$$P = (T_1 – T_2) \times v \ (1646.5 – 608.2) \times 11.31 \text{ N–m/s}$$
$$= 11,740 \text{ W} = 11.74 \text{ kW}$$

(ii) Power transmitted when the centrifugal tension is neglected

We know that tension in the tight side (without centrifugal tension), $T_1 = 1800$ N

Therefore, $T_2 = 1800/2.707 = 665$ N and power transmitted,

$$P = (T_2 - T_1) \, v = (1800 - 665) \, 11.31. \text{ N–m/s}$$
$$= 12,840 \text{ W} = 12.84 \text{ kW}$$

Problem 1.10 A flat belt is required to transmit 3.5 kW from a pulley of 1.5 m effective diameter running at 300 rpm. The angle of contact is spread over 11/24 of the circumference and the coefficient of friction between the belt and pulley surface is 0.3. Taking centrifugal tension into account, determine the width of the belt. Take belt thickness as 9.5 mm, density as 1.1 Mg/m³ and permissible stress as 2.5 N/mm².

Solution: Given, P = 3.5 kW; d = 1.5 m; N = 300 rpm; $\theta = 2\pi \times 11/24$ = 288 rad; μ = 0.3; t = 9.5 mm = 0.0095 m; ρ = 1.1 Mg/m³ = 1100 kg/m³; f = 2.5 N/mm².

Let b   = Width of the belt in mm,

T_1 = Tension on the tight side of the belt, and

T_2 = Tension on the slack aide of the belt.

We know that velocity of the belt,

$$v = \pi \times 1.5 \times 300/60 = 23.56 \text{ m/s and}$$

power transmitted (P),

$$35 = (T_1 - T_2) \, v = (T_1 - T_2) \times 23.56$$

Therefore,

$$(T_1 - T_2) = 35/23.56 = 1.486 \text{ kN} = 1486 \text{ N} \qquad \dots \text{(a)}$$

We also know that

$$2.3 \log (T_1/T_2) = \mu.\theta = 0.3 \times 2.88$$
$$= 0.864 \log (T_1/T_2) = 0.864/2.3 = 0.3757$$

Therefore, $(T_1/T_2) = 2.375$ (Taking antilog of 0.3757) or

$$T_1 = 2.375 \, T_2$$

Substituting the value of T_1 in equation (a),

$$2.375 \, T_2 - T_2 = 1486$$

Therefore, $T_2 = 1486/1.375 = 1081$ kN and $T_1 = 2.375 \times 1081 = 2567$ kN

We know that maximum tension in the belt,

$$T = fbt = 2.5 \times b \times 9.5 = 23.75b \text{ N and}$$

mass of the belt per metre length,

m = Area × Length × Density

(0.00b × 0.0095) × (1) × (1100) = 0.010 45 b kg

Therefore, centrifugal tension,

T_c = m.v² = 0.010 45b (23.56)² = 5.8b N and

tension on the tight side of the belt (T_1)

2567 = T – T_c = 2.375b – 58b = 17.95b

Therefore, b 2567/17.95 = 143 mm, say 150 mm

Problem 1.11: A belt 100 mm wide and 10 mm thick is transmitting power at 1200 m/min. The net driving tension is 18 times the tension on the slack side. If the safe permissible stress on belt section is 1.8 N/mm², calculate the power that can be transmitted at this speed. Assume mass density of the leather as 1 t/m³.

Also calculate the absolute maximum power that can be transmitted by this belt and the speed at which this can be transmitted.

Solution: Given, b =100 mm; t = –10 mm; v =1200 mm/min = 20 m/s; $(T_1 – T_2)$ = 1.8 T_2; f = 1.8 N/mm²; ρ = 1 t/m³

Power transmitted by the belt

We know that maximum tension in the belt,

T = fbt = 1.8 × 100 × 10 = 1800 N and

mass of the belt per metre length,

m = Area × Length × Density

(0.1 × 0.001) × (1) × (1000) 1 kg/m

Therefore, centrifugal tension,

T_c = mv² = 1(20)² = 400 N

and tension in the tight side,

T_1 = T – T_c =1800 – 400 = 1400 N

Now 1400 – T_2 = 1.8 T_2 (Given T_1 – T_2 = 1.8 T_2) T_2 = T_1/2.8 = 1400/2.8 = 500 N and power transmitted by the belt,

P = $(T_1 – T_2)$ v = (1400 – 500) 20 N–m/s

18,000 W = 1.8 kW

Speed at which absolute maximum power can be transmitted

We know that speed of the belt, at which maximum can be transmitted,

$$v = \sqrt{T/3m} = \sqrt{1800/3 \times 1} = 24.5 \text{ m/s}$$

Absolute maximum power that can be transmitted by the belt

We know that for maximum power, the centrifugal tension, $T_c = T/3 = 1800/3 = 600$ N Therefore, tension on the tight side, of the belt, $T_1 = T - Tc = 1800 - 600 = 1200$ N and tension on the slack side of the belt, $T_2 = T_1/2.8 = 1200/2.8 = 428.6$ N

Problem 1.12: Two parallel shafts whose centre lines are 4.8 m apart are connected by an open belt drive. The diameter of the larger pulley is 1.5 m and that of the smaller pulley is 1 m. The initial tension in the belt, when stationary, is 30 kN. The mass of the material is 1.5 kg/m length and the coefficient of friction between the belt and the pulley is 0.3. Calculate the power transmitted, when the smaller pulley rotates at 400 rpm.

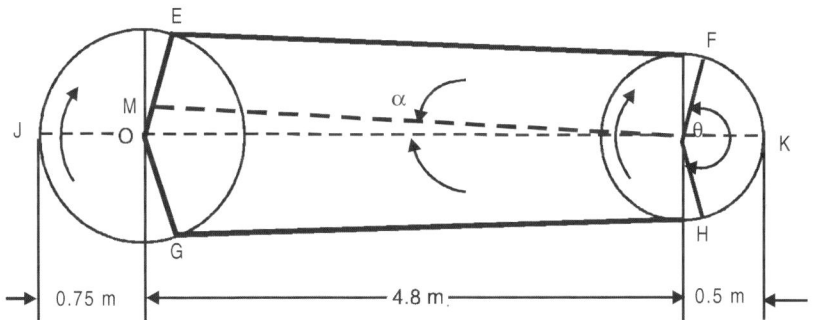

Solution: Given, l = 4.8 m; d_1= 1.5 m r_1= 0.75 m; d_2 = 1 m; or r_2 = 0.5 m; T_0 = 3 kN; m = 1.5 kg/m; μ = 0.3; N_2 = 400 rpm. Let T_1 = Tension in the tight side of the belt, and T_2 = Tension in the slack side of the belt. We know that velocity of the belt,

$$v = \pi \, d_2 \, N \div 60 = (\pi \times 1 \times 400) \div 60 = 20.94 \text{ m/s}$$

and initial tension in the belt when stationary (T_0)

$$3 = (T_1 + T_2) \div 2 \text{ or } T_1 + T_2 = 6 \text{ kN} \qquad \dots \text{(a)}$$

Now for an open belt drive,

$$\text{Sin } ά = (r_1 - r_2) \div l = (0.75 - 0.5) \div 4.8 = 0.0521 \text{ or } ά = 3°$$

Therefore, Angle of lap for the smaller pulley,

$\theta = 180° - 2\acute{\alpha} = 180° - (2 \times 3) = 174° = \pi \times 174/180 = 3.04$ rad.

We also know that

$2.3 \log (T_1/T_2) = \mu\theta = 0.3 \times 3.04 = 0.912$

Therefore,

$\log (T_1/T_2) = 0.912 / 2.3$

$= 0.3965$ or $T_1/T_2 = 2.49$ or $T_1 = 2.49 \ T_2$

(Taking antilog of 0.3965)

Substituting the value of T_1 in equation (a)

$2.49 \ T_2 + T_2 = 6$

Therefore, $T_2 = 6/3.49 = 1.72$ kN and $T_1 = 2.49 \ T_2 = 2.49 \times 1.72 = 4.28$ kN
Therefore, Power transmitted by the belt,

$P = (T_1 - T_2) \ v = (4.28 - 1.72) \ 20.94 \ \text{kN–m/s} = 53.6 \ \text{kW}$

Problem 1.13: Find the power transmitted by a rope drive, from the following data:

Angle of contact 180°; Pulley groove angle 60°; Coefficient of friction = –0.2; Mass of rope 0.4 kg/m length; Permissible tension 1.5 kN; Velocity of rope 1.5 m/s.

Solution: Given, $\theta = 180° = 3.142$ rad; $2\alpha = 60°$ or $\alpha = 30°$; $\mu = 0.2$; m = 0.4 kg/m; T = 1.5 kN; $v = 15$ m/s.
We know that the centrifugal tension,

$T_c = m.v^2 = 0.4(15)^2 = 90\text{N}$

Therefore, $T_1 = T - T_c = 1500 - 90 = 1410$ N and

$2.3 \log (T_1/T_2) = \mu.\theta \ \text{cosec} \ \alpha$

$= 0.2 \times 3.142 \ \text{cosec} \ 30°$

$= 0.2 \times 3.142 \times 2.0 = 1.257$

Therefore, $\log (T_1/T_2) = 1.257 \times 2.3 = 0.5465$ or

$1410 \div T_2 = 3.52$ (Taking antilog of 0.5465)

Therefore, $T_2 = 1410/3.52 = 400$ N
We know that power transmitted by the rope drive,

$P = (T_1 - T_2)v = (1410 - 400) \times 1.5 \ \text{N–m/s},$

$15,150 \ \text{W} = 15.15 \ \text{kW}$

Problem 1.14: A rope pulley with 5 ropes and surface speed of 1000 m/min transmits 100 hp. Find the tensions on the tight side and slack side, if the angle of lap is 130°, and the angle between the sides of the groove is 45°. Assume μ = 0.3.

Power transmitted per rope,

$$hp = 100 \div 5 = 20 \text{ hp}$$

In case of rope pulleys,

$$\log T_1 \div T_2 = 0.4343 \times [\mu\theta/Sin\alpha] \quad \alpha = 45° \div 2$$
$$= 22.5° \, [2\alpha = 45°]$$
$$\theta = [\pi \div 180] \times 130 = 2.27 \text{ radian}$$

Therefore,

$$\log (T_1 \div T_2) = 0.4343 \times (0.3 \times 2.27) \div \sin 22.5 = 0.77,$$
$$(T_1 \div T_2) = 5.888$$
$$hp = [(T_1 - T_2) \times v] \div 4500 \text{ or}$$
$$20 = [(T_1 - T_2) \times 1000] \div 4500 \, (T_1 - T_2)$$
$$= (20 \times 4500) \div 1000$$
$$= 90 \text{ kgf} = 5.888 \, T_2 - T_2 \text{ or}$$
$$T_2 = 18.4 \text{ kgf or } T_1 = 5.888$$
$$T_2 = 5.888 \times 18.4 = 108.3 \text{ kgf}$$

Problem 1.15: A rope drive is required to transmit 1 MW from a pulley of 1 m diameter running at 450 rpm. The safe pull in each rope is 224 kN and the rope has mass of 1 kg/m. The angle of lap and the groove angle are 150° and 45°, respectively. Find the number of ropes required for the drive, if the coefficient friction between the rope and the pulley is 0.3.

Solution: Given, P = 1 MW = 1000 kW; d = 1 m; N = 450 rpm; T = 22.5 kN = 2250 N; m = 1 kg/m; θ = 150° = 150 π/180 = 2.62 rad; 2α = 45° or α = 22.5°; μ = 0.3.

We know that velocity of the ropes,

$$v = \pi \times 1 \times 450/60 = 23.56 \text{ m/s}$$

Centrifugal tension,

$$T_c = m.v^2 = 1(23.56)^2 = 555 \text{ N}$$
$$T_1 = T - T_c = 2250 - 555 = 1695 \text{ N and}$$
$$2.3 \log (T_1/T_2) = \mu\theta \text{ cosec } \alpha = 0.3 \times 2.62 \text{ cosec } 22.5°$$
$$= 0.3 \times 2.62 \times 2.613 = 2.054$$
$$\log (T_1/T_2) = 2.054/2.3 = 0.8930$$

$$1695 \div T_2 = 7.816 \qquad \text{(taking anti-log of 0.8930)}$$

Therefore,

$$T_2 = 1695/7.816 = 217 \text{ N and}$$

Power transmitted by one rope

$$= (T_1 - T_2) \, v = (1695 - 217) \, 23.56 \text{ N–m/s}$$
$$= 34820 \text{ kW} = 34.82 \text{ kW}$$

Therefore, no. of ropes

$$= \frac{\text{Total power to be transmitted}}{\text{Power transmitted by one rope}} = \frac{1000}{34.82}$$
$$= 28.7, \text{ say } 29$$

Normal belt life requirements	kh
Industrial –	12–26
• Equipment with long operating hours or continuous operations like fans, pumps, conveyors	7.5–20
• Hand tools, office equipment	6–12
• Equipment with intermittent or occasional operation	
Agricultural –	
• Stationary equipment with long operating hours or continuous operations like fans, pumps, conveyors	6–12
• Stationary equipment with intermittent or occasional operation	2–6
• Mobile equipment like harvesters, sowing machines, manure spreaders, hay balers	0.5–1
Automotive –	
• Passenger cars, vans	1–3
• Lorries, buses, tractors, road construction machines	5–10
Home Appliances –	
• Heating, ventilation, air conditioning	5–10
• Washing machines, tumbler dryers, dish washers	1.5–2
• Sewing machines, lawn mowers, hand tools	0.2–1

2.1 Introduction

Gears are the positive means of transmitting power between two shafts. They are more widely used than belts and chains in power transmission. The gears can be straight tooted, spiral/helical toothed and bevel toothed. Straight toothed gears are used to transmit power between horizontal shafts, whereas bevel gears are used to transmit power between perpendicular shafts. Nowadays helical gears are being used in place of straight tooth gears, since they can rotate at higher speeds without getting worn out. Gears are widely used in spinning as well as weaving machines. In weaving machines they are used for take up, and let off motions. Bevel gears are used in the speed frame builder motion, and also in sizing machines. Gears are considered to be positive drives as there is no slippage.

2.2 Classification of gears

Gears are classified into the following types

(a) Simple gears
(b) Compound gears
(c) Epicyclic gears – simple and compound (sun and planet wheel)

2.3 Important terminologies

Addendum

It is the height by which a tooth projects beyond the pitch circle or pitch line.

Base diameter

It is the diameter of the base cylinder from which the involute portion of a tooth profile is generated.

Backlash

It is the amount by which the width of a tooth a tooth space exceeds the thickness of the engaging tooth on the pitch circles.

Bore length

It is the total length through a gear, socket or coupling bore.

Circular pitch

It is the distance along the pitch circle or pitch line between corresponding profiles of adjacent teeth.

Circular thickness

It is the length of arc between the two sides of a gear tooth on the pitch circle unless otherwise specified.

Clearance operating

It is the amount by which the dedendum in a given gear exceeds the addendum of its mating gear.

Contact ratio

In general it is the number of angular pitches through which a tooth surface rotates from the beginning to the end of contact.

Dedundum

It is the depth of a tooth space below the pitch line. It is normally greater than the addendum of the mating gear to provide clearance.

Diametral pitch

It is the ratio of the number of teeth to the pitch diameter.

Face width

It is the length of the teeth in an axial plane.

Fillet radius

It is the radius of the fillet curve at the base of the gear tooth.

Full depth teeth

These are ones in which the working depth equals 2.000 divided by the normal diametrical pitch.

Gear

It is a machine part with gear teeth. When two gears run together, the one with the larger number of teeth is called the gear.

Hub diameter

It is the outside diameter of a gear, sprocket, or coupling hub.

Hub projection

It is the distance the hub extends beyond the gear face.

Involute teeth (spur gears, helical gears, and worms)

These are ones in which the active portion of the profile in the transverse plane is the involute of a circle.

Long and short addendum teeth

These are engaging gears (on a standard designed centre distance) one of which has a long addendum and the other has a short addendum.

Keyway

It is the machined groove running the length of the bore. A similar groove is machined in the shaft and a key fits into this opening.

Normal diametrical pitch

It is the value of the diametrical pitch as calculated in the normal plane of a helical gear or worm.

Normal plane

It is the plane normal to the tooth surface at a pitch point and perpendicular to the pitch plane. For a helical gear this plane can be normal to one tooth at a point laying in the plane surface. At such a point, the normal plane contains the line normal to the tooth surface and this is normal to the pitch circle.

Normal pressure angle

It is a normal plane of helical tooth.

Outside diameter

It is the diameter of the addendum (outside circle) circle.

Pitch circle

It is the circle derived from a number of teeth and a specified diametrical or circular pitch. Circle on which spacing or tooth profiles is established and from which the tooth proportions are constructed.

Pitch cylinder

It is the cylinder of diameter equal to the pitch circle.

Pinion

It is a machine part with gear teeth. When two gears run together, the one with the smaller number of teeth is called the pinion.

Pitch diameter

It is the diameter of the pitch circle. In parallel shaft gears, the pitch diameters can be determined directly from the centre distance and the number of teeth.

Pressure angle

It is the angle at a pitch point between the line of pressure which is normal to the tooth surface, and the plane tangent to the pitch surface. In involute teeth, pressure angle is often described also as the angle between the line of action and the line tangent to the pitch circle. Standard pressure angles are established in connection with the standard gear tooth proportions.

Root diameter

It is the diameter at the base of the tooth space.

Pressure angle operating

It is determined by the centre distance at which the gears operate. It is the pressure angle at the operating pitch diameter.

Tip relief

It is an arbitrary modification of a tooth profile whereby a small amount of material is removed near the tip of the gear tooth.

Undercut

It is a condition in generated gear teeth when any part of the fillet curve lies inside a line drawn tangent to the working profile at its point of juncture with the fillet.

Whole depth

It is the total depth of a tooth space, equal to addendum plus dedundum, equal to the working depth plus variance.

Working depth

It is the depth of engagement of two gears; that is, the sum of their addendums.

2.4   Simple gears

A simple gear train uses two gears, which may be of different sizes. If one of these gears is attached to a motor or a crank then it is called the driver gear. The gear that is turned by the driver gear is called the driven gear.

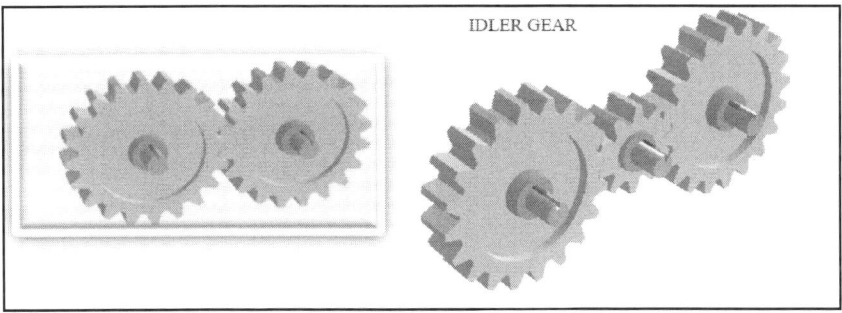

2.1 Simple gear train.

When a simple gear train has three meshed gears, the intermediate gear between the driver and driven gear is called an idler gear. An idler gear does not affect the gear ratio (velocity ratio) between the driver gear and the driven gear. It is mainly used to bridge the gap between the driver and driven wheels which are a distance apart and may also be used to get same direction of rotation of driven as that of driver.

2.5 Compound gears

Compound gear trains involve several pairs of meshing gears. They are used where large speed changes are required or to get different outputs

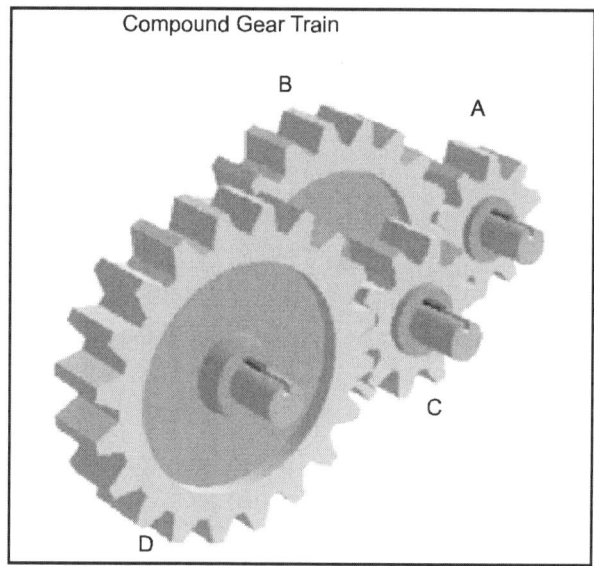

2.2 Compound gear train.

moving at different speeds. Gear ratios (or velocity ratios, VR) are calculated using the same principle as for simple gear trains, i.e. VR = number of teeth on the driver gear divided by the number of teeth on the driven gear.

However, the velocity ratio for each pair of gears must then be multiplied together to calculate the total velocity ratio of the gear train:

$$\text{Total VR} = \text{VR}_1 \times \text{VR}_2 \times \text{VR}_3 \times \text{VR}_4, \text{ etc.}$$

$$\text{Gear ratio} = \frac{\text{no. of teeth on B}}{\text{no. of teeth on A}} \times \frac{\text{no. of teeth on D}}{\text{no. of teeth on C}}$$

2.6   Compound epicyclic gear train (sun and planet wheels)

A compound epicyclic gear train consists of three toothed wheels known as the sun wheel (S), the planet wheel (P) and annular wheel (A) as shown in Fig. 2.3. The axes of sun wheel and planet wheel are connected by an arm C by pin connections. The planet wheel meshes with the sun wheel as well as the annular wheel.

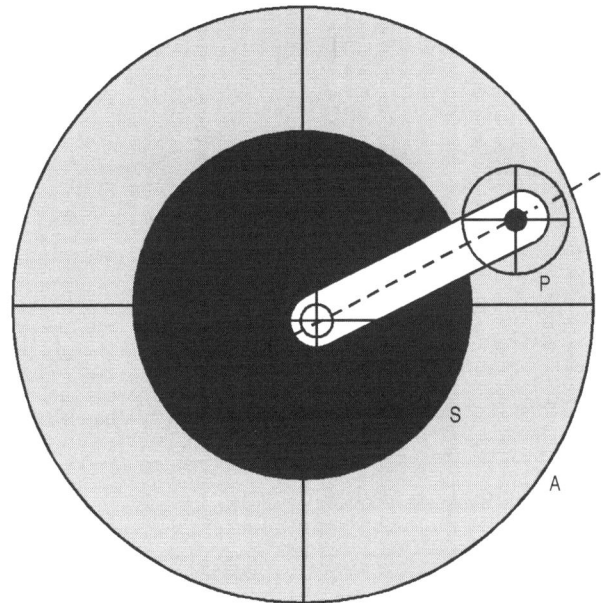

2.3 A compound epicyclic gear train.

It may be noted, that the planet wheel (P) rotates about its own axis and at the same time it is carried round the sun wheel (S) by the arm C. A little consideration will show that, when the sun wheel (S) is fixed the annular wheel (A) provides the drive. But when the annular wheel (A) is fixed the sun wheel (S) provides the drive. In both the cases, the arm C acts as a follower.

Let T_A = No. of teeth on the annular wheel A,
        N_A = Speed of the annular wheel A,
        T_s, N_s = Corresponding values for the sun wheel (S), and
        T_s, N_s = Corresponding values for the planet wheel (P)

Now the velocity ratio of a compound epicyclic gear train may be obtained by preparing a table of motions as usual.

Example 1: An epicyclic gear consists of three wheels A, B and C as shown in Fig. 2.4. The wheel A has 72 internal teeth, C has 32 external teeth. The wheel B gears with both the wheels A and C and is carried on an arm D, which rotates about the centre of wheel A at 18 rpm. Determine the speed of the wheels B and C, when the wheel A is fixed.

Solution: Given, T_A = 72; T_C = "32; N_D = 18 rpm.
First of all, prepare the table of motions as given below:

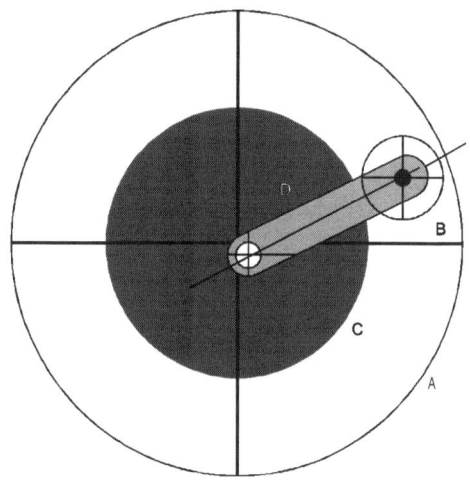

2.4 Epicyclic gear train.

Step no.	Conditions of motion	Revolution of			
		Arm	Wheel C	Wheel B	Wheel A
1.	Arm fixed; wheel C rotates through +1 revolution	0	+ 1	$-T_C/T_B$	$-T_C/T_B \times T_B/T_A$
2.	Arm fixed; wheel C rotates through + x revolution	0	+ x	$-x \times T_C/T_B$	$-x \times T_C/T_A$
3.	Add +y revolutions to all elements	+ y	+ y	+ y	+ y
4.	Total motion 2 + 3	+ y	x + y	$(y-x) \times T_C/T_B$	$(y-x) \times T_C/T_A$

For speed wheel C,

We know that the speed of the arm, y = 18 rpm.

Since the wheel A is fixed, $\therefore (y-x) \times T_C/T_A = 0$

or $18 - x \times \dfrac{32}{72} = 0$; $18 \times 32 - 32x = 0$

$\therefore x = 18$rpm and speed of wheel C,

$N_C = x + y = 18 + 18 = 36$ rpm.

For speed of wheel B,

Let d_A, d_B and d_C be the pitch circle diameters of wheels A, B and C, respectively. From the geometry of Fig. 2.4, we find that

$$d_B + \frac{d_C}{2} = \frac{d_A}{2} \quad \text{or} \quad 2d_B + d_C = d_A$$

Since the number of teeth are proportional to their diameters, therefore

$$2T_B + T_C = T_A \quad \text{or} \quad 2T_B + 32 = 72$$

$$\therefore T_B = 20; \text{ and speed of wheel B, } N_B = y - x\frac{T_C}{T_B} = 18 - 40.5 \times \frac{32}{20}$$

$$= -46.8 \text{ rpm}$$

Example 2: An epicyclic gear train, as shown in Fig. 2.5, has a sun wheel **S** of 30 teeth and two planet wheels **P, P** of 50 teeth. The planet wheels mesh with the internal teeth of a fixed annular **A**. The driving shaft carrying the sun wheel transmits 4 kW at 300 rpm. The driven shaft is connected to an arm, which carries planet wheels. Determine the speed of the driven shaft and the torque transmitted, if the overall efficiency is 95%.

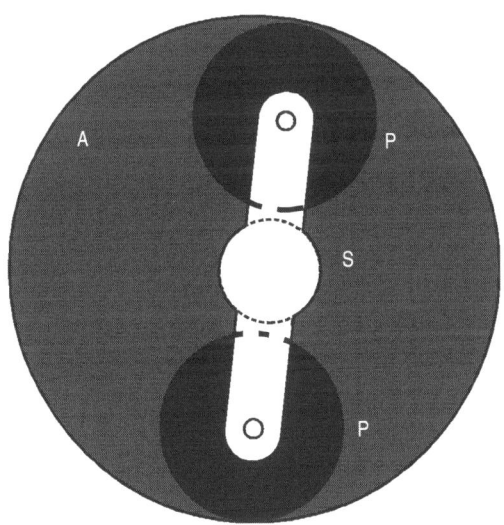

2.5 Epicyclic gear train.

Solution: Given, $T_s = 30$; $T_p = 50$; P = 4 kW = 4000 W; speed of driving shaft = 300 rpm; ç = 95% = 0.95;

∴ Power transmitted by the driven shaft = 4000 × 0.95 = 3800 W.

First of all prepare the table of motions as given below:

Step no.	Conditions of motion	Revolution of			
		Arm	Wheel A	Wheel P	Wheel S
1.	Arm fixed; wheel A rotates through + 1 revs.	0	+ 1	$+\dfrac{T_A}{T_P}$	$-\dfrac{T_A}{T_P}\times\dfrac{T_P}{T_S}$ $=-\dfrac{T_A}{T_S}$
2.	Arm fixed; wheel A rotates through + x revs.	0	+ x	$+x\times\dfrac{T_A}{T_P}$	$-x\times\dfrac{T_A}{T_S}$
3.	Add + y revs. to all elements	+ y	+ y	+ y	+ y
4.	Total motion	+ y	x + y	$y+x\times\dfrac{T_A}{T_P}$	$y-x\times\dfrac{T_A}{T_S}$

Speed of the driven shaft

Let d_A, d_P and d_S be the pitch circle diameters of the wheel A, P and S, respectively.

From the geometry of Fig. 2.4 we find

$$\frac{d_A}{2}=\frac{d_S}{2}+d_P \text{ or } d_A=d_S+2d_P$$

Since the number of teeth are proportional to their diameters, therefore

$$T_A = T_S + 2T_P = 30 + (2\times50) = 130$$

and as the wheel *A* is fixed, therefore

$$x + y = 0 \text{ or } x = -y \qquad\qquad … (a)$$

Moreover, as the sun wheel S rotates at 300 rpm. Therefore,

$$300 = y - x \times\frac{T_A}{T_S} = y - x \times \frac{130}{30} = y - \frac{13x}{3}$$

$$300 = y - \frac{13}{3}(-y) = y(1+\frac{13}{3}) = \frac{16y}{3}$$

$$\therefore \qquad y = \frac{300\times3}{16} = \frac{900}{16} \text{ and } x = -y = -\frac{900}{16}$$

Therefore speed of the driven shaft,

N = Speed of arm

$$= y = \frac{900}{16} = 56.25 \text{ rpm}$$

Torque transmitted by the driven shaft
Let T = Torque transmitted by the shaft
We know that power transmitted by the shaft,

$$300 = 2 \partial NT = 2 \partial \times 56.25 \times T = 353.4 \times T$$

∴ $$T = \frac{3800}{353.4} = 10.72 N - m$$

2.7   Epicyclic gear train with bevel wheels

The problems on epicyclic gear train, with bevel wheels, may be solved exactly in the same manner as in the case of epicyclic gear train with spur gears. The only important point, in such problems, is that the direction of motion of the two bevel (i.e. slanting) wheels (connected by an intermediate spindle) is unlike.

Example: An epicyclic gear train consists of bevel wheel as shown in Fig. The wheel A, which is keyed to driving shaft x, has 40 teeth and meshes with the wheel B (50 teeth) which in turn meshes with the wheel C having 20 teeth. The wheel C is keyed driven shaft Y. The wheel B turns freely on the arm which is rigidly attached to the hollow sleeve. The hollow sleeve is riding freely loose on the axis of the shafts X and Y.

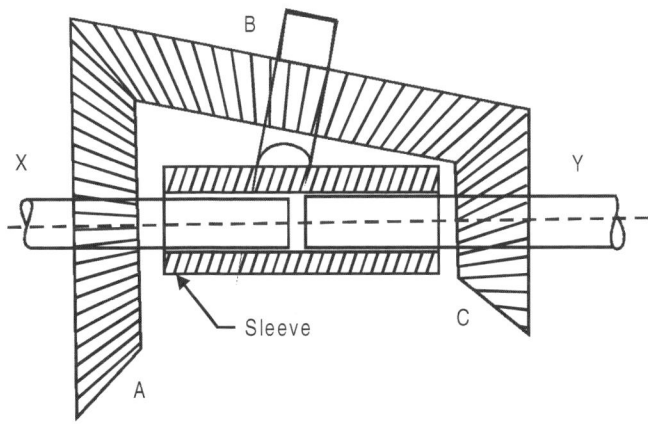

If the driving shaft rotates at 50 rpm anticlockwise and the arm rotates at 100 rpm clockwise, determine speed of the driven shaft.

Solution: Given, $T_A = 40$; T_B –50; $T_C = -20$; Speed of driving shaft = –50 rpm (anticlockwise); Speed of arm = 100 rpm (clockwise).

First of all prepare the table of motions as below.

Step no.	Conditions of motion	Revolution of			
		Arm	Wheel A	Wheel P	Wheel S
1.	Arm fixed; wheel A rotates through + 1 rev.	0	+ 1	$+\dfrac{T_A}{T_B}$	$-\dfrac{T_A}{T_B}\times\dfrac{T_B}{T_C}=-\dfrac{T_A}{T_C}$
2.	Arm fixed; wheel A rotates through + x revs.	0	+ x	$+x\times\dfrac{T_A}{T_B}$	$-x\times\dfrac{T_A}{T_C}$
3.	Add + y revs. to all elements	+ y	+ y	+ y	+ y
4.	Total motion	+ y	x + y	$y+x\times\dfrac{T_A}{T_B}$	$y-x\times\dfrac{T_A}{T_B}$

Since the speed of the arm is 100 rpm (clockwise), therefore y = +100
Moreover, as the speed of the driving shaft is 50 rpm (anticlockwise) therefore

$$x + y = -50 \text{ or } x = -50 - y = -50 - 100 = -150$$

Speed of the driving shaft = Speed of wheel C

$$y - x \times \frac{T_A}{T_C} = 100 + 150 \times \frac{40}{20} = 400 \text{ rpm}$$

2.8 PIV gears

Positively infinitely variable drives are nowadays widely used. They can vary the speed in even small increments. The modern concept of the PIV is the CVT (continuously variable transmission). The continuously variable transmission is a system that makes it possible to vary stepless the transmission ratio without interruption of the torque flow. Therefore an infinite number of ratios (between a minimum and maximum value) are possible. The key lies in its simple yet effective belt-and-pulley design. The CVT works with an all-metal chain that runs between cone-shaped curved pulleys. The transmission ratio between the engine and drive wheels changes in a smooth manner in relation to the variable axial gap between the pulleys. The gap defines the possible chain radii on the pulleys. An infinite number of transmission ratios allows the engine to always run at its optimum speed. The 'stepless' nature of its design is CVT's biggest draw for automotive engineers. Because of this, a CVT can work to keep the engine in its optimum power range, thereby increasing efficiency and gas mileage. The PIV Drives CVT is especially designed for heavy duty and high torque applications in the field of construction and off-road where a robust design is needed. The CVT's efficiency is optimized for all ratios. A CVT can convert every point on the engine's operating curve. An optimal drive-line management is attainable because every kind of operating strategy can be considered.

The operational principle of the CVT drives relies upon frictional engagement of the chain's rocker pins onto the surfaces of the input pulley and output pulley. Contact pressure from the discs onto the chain is applied through e.g. oil pressure in the cylinders of the movable discs. The torque sensor, a special feature of the PIV clamping system, is located in the power train. It ensures that there is sufficient pressure to prevent the chain from slipping. A superimposed control system is used for setting the transmission ratio.

Advantages

- High reliability and endurance life through metal torque transmitting components with lubrication
- Low noise emission through a special chain design (patented)
- High precision of output speed combined with high-speed ratio changing
- Shifting while idling is possible
- No fixed relation between engine speed and machine speed
- Optimization of working processes with engines having performance charts with characteristic fields; such as specific fuel consumption
- Drive-line management with one engine and several power consuming systems; such as PTOs
- High efficiency due to optimized clamping systems
- High power performance through computer-optimized chains and sheaves.

2.9 Worm and worm wheel

In this case a single- or double-threaded worm drives a worm wheel. This is used for slow speeds.

Textile applications of gears

1. Simple gears are widely used in textile machines from spinning to weaving and wet processing machines.

2. Compound gears are used in weaving machines, such as those in take up mechanism.
3. Epicyclic gears are used in speed frames for building motion, and in Roper let off mechanism in the loom.
4. Worm and worm wheel is used in the continuous positive take up mechanism.

3

Design of cone pulleys in textile machines

3.1   Introduction

Cone drums/pulleys are used in spinning in the areas of blow room and roving frames. In the case of blow room, they are used in scutcher for regulation of the feed of cotton. The feed material in scutcher has irregularities such thick and thin places. So it is necessary to maintain a uniform weight/unit length of the feed material to scutcher. For doing this, the feed material will have to be fed at different speeds as per the thickness. The cone drums/pulleys help in this regard and thereby ensure a uniform amount of feed material at a given unit of time and thereby maintain constant uniformity of material. In the case of the speed frames, the cones enable to give differential speeds to bobbin. In order to maintain the same amount length/unit weight of the roving, the bobbin will have to be driven at different speeds during the bobbin build. The speed has to be gradually reduced with increasing bobbin diameter, so as to maintain the same amount of material in a given unit time. This chapter discusses in detail about the design aspects of the cone pulleys used in both cases.

3.2   Design of cone pulleys for piano feed regulation of web in blow room

In piano feed regulation of a scutcher, several pedals below feed roller moves up and down independently, depending on the localized material thickness variations. Through links and levers, these movements are mechanically integrated. The integrated output is used to move the belt on cone drums, varying the rpm of top cone pulley.

The output from the top cone is transmitted to feed roller, through gears, adjusting its speed corresponding to average thickness of material in between pedals and feed roller. When the average thickness of material between the pedal and the feed roller is high, the speed of the feed roller has to be reduced and vice versa.

The following assumptions are useful in the design of the cones:

- Density of web is constant, i.e. mass flow rate of web is proportional to its thickness.

- Sum of the top and bottom cone diameters for any cone belt position (or web thickness) is constant.
- Specification for cones dimensions, viz., maximum and minimum diameters, and length.
- Rotational rates of driving cone (bottom cone).

$$\text{The mass of a unit length of a web (m) is } m = kt \quad \ldots (3.0)$$

where   k = areal density of web in g/cm^2, which is constant.

       t = Web thickness in cm.

Therefore, the mass flow rate of the web (M) in g/min is

$$M = kt\,(\pi d_f n_f) \qquad \ldots (3.1)$$

where d_f = Diameter of feed roller in cm.

       n_f = rpm of feed roller.

When the gear transmission ratio from top cone to feed roller is 'e', then

$$Nf = eN_2 \qquad \ldots (3.2)$$

where N_2 = rpm of driven cone (top cone)

$$M = \pi.k.t.d_f\,.e.N_2 \qquad \ldots (3.3)$$

Since M, k, π, d_f, and e are constant, under a given practical condition, then, N_2 is proportional to 1/t

$$\text{Also, } N_2 = (D_1 N_1)/D_2$$

where l_3 = Bottom cone diameter in cm

      132 = Top cone diameter in cm

      N_1 = bottom cone rpm

Therefore, 1/t is proportional to $D_1 N_1/D_2$. Since N_1, the rpm of driving cone pulley is kept constant, (1/t) is proportional to N_2 or

From the above expressions

$$N_2/N_1 = = [M\,/(\pi kdf.e\;N_1)]/t = \text{constant} / t \qquad \ldots (3.4)$$

By assuming various thickness values for the web passing between feed roller and pedals, the ratio (N_2/N_1) or D_1/D_2 can be found out. From this ratios, and D_2 values can be computed considering that $D_1 + D_2 = X$, a constant. This is given in the Table 3.1.

Construct the cones similar to the procedure followed for the designing of the cone pulleys discussed for the speed frame. In practice, the web thickness variation should be considered as infinitesimal to get smooth profile for the cone surfaces.

Table 3.1 Calculation of the diameters of driving and driven pulleys for piano-feed regulation.

Web thickness (cm)	$(1/t) = (D_1/D_2)$	+	D_1	D_2
0.5	1/0.5	×	calculate	calculate
0.6	1/0.6	×		
0.7	1/0.7	×		
0.8	1/0.8	×		
0.9	1/0.9	×		
1.0 (normal)	1/1.0	×		
1.1	1/1.1	×		
1.2	1/1.2	×		
1.3 and so on	1/1.3 and so on	×		

3.3   Design of cone pulleys for speed frame

Bobbin speed is a function of bobbin diameter. The reduction in bobbin speed is effected through cone pulleys. After the completion of winding of each layer, the builder mechanism shifts the belt on the cone pulleys, in such a way that the speed of bottom cone pulley is reduced. From the bottom cone pulley, drive is transmitted to the differential gearing via gear trains. This speed is superimposed on the fixed speed of the main shaft or arm of the differential gearing. The output from the epicyclic gearing is transmitted to the bobbins though gears.

3.3.1   Design aspects

To have a simplified approach in designing the cone pulleys, certain assumptions have to be made. Boundary conditions for the design can be set based on space consideration, dimensions of related elements (gear trains). Based on the space consideration and for ease of design calculations, the dimensions of the following can be selected.

- Maximum and minimum diameter of cone pulleys.
- Sum of top and bottom cone pulley diameters for any cone belt position is constant.
- Length of cone pulleys.

Based on the dimensions of the following elements, a design criterion has to be set.

- Width of cone belt.
- Minimum and maximum possible bobbin diameters (bare and full bobbin).

The consideration of the following points could simplify the design approach further.

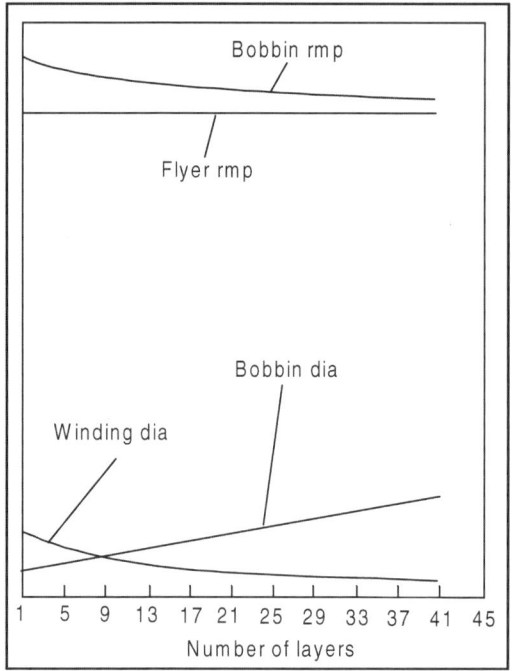

3.1 Bobbin diameter and speeds as a function of number of layers.

Where, L = gear train ratio from output wheel '6' of differential gear box to bobbin.

e = epicyclic gear train ratio;

K = gear train ratio from bottom cone to gear '1'.

= Winding rpm.

4. Then top cone rpm is

$$N_1 = FA \qquad \qquad \dots (3.5)$$

Where, F = rpm of main shaft.

A = gear train ratio from main shaft to top cone.

The rpm of top and bottom cones are plotted in Fig. 3.2.

5. Ratio of top to bottom cone diameters $(d_1/d_2) = (N_2/N_1)$ for various number of layers. This is plotted in Fig. 3.3. (Note: Plotting of this ratio is not essential, and only the values are needed for design).

6. From this, d_1 and d_2 values for various bobbin diameters can be calculated from the expression

$$d_1 + d_2 = \text{constant}. \qquad \qquad \dots (3.6)$$

7. Mark the r_1 and r_2 values (radii of cones) from the axis of both cones for infinitesimal shift of belt positions.

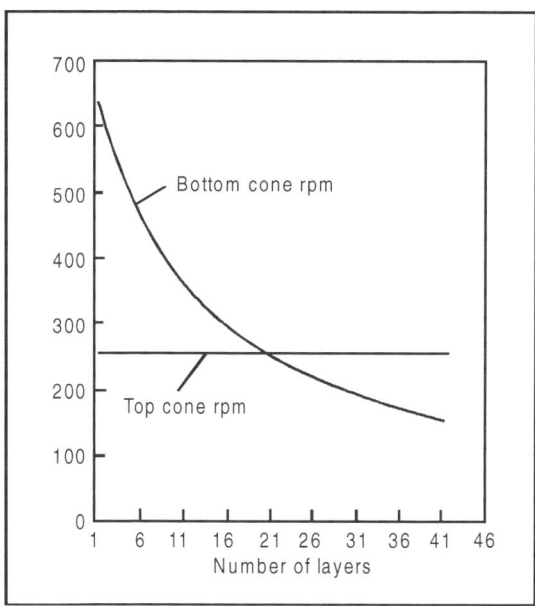

3.2 Rpm of top and bottom cones.

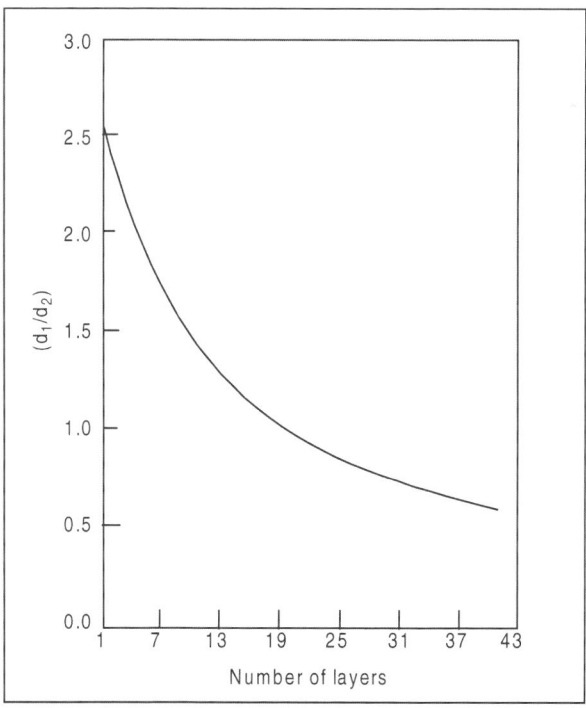

3.3 Ratio of top to bottom cone diameters.

8. Draw smooth curve tracing the points as shown in Fig. 3.4.

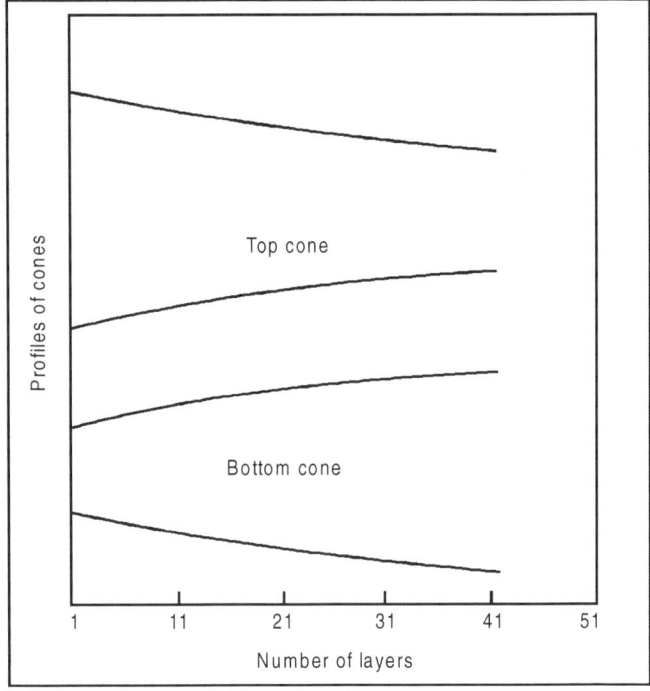

3.4 Profiles of cones.

3.4   The profile of the cone drums

Let the front rollers deliver a length L per minute. Let the bobbin diameter at any instant be B. Let the top-cone-drum diameter be $\varnothing_T$ and the bottom-cone-drum diameter $\varnothing_B$.

At any instant, the excess rev/mm of the bobbin over the spindle is $L/\pi B$

Let the top-cone-drum speed be R rev/mm. Then, the bottom-cone-drum speed is

$$R \times \varnothing_T / \varnothing_B$$

The bottom-cone-drum speed multiplied by the gearing constant of the drive from the bottom cone drum to the bobbin equals the excess rev/mm. Let this constant be k_1, we then have

$$R \times \varnothing_T / \varnothing_B \times k_1 = L/\pi B$$

Since R, L, π, and k_1 are all constants, we can rewrite the last line in the form:

$$\varnothing_T \,/\, \varnothing_B = K/B, \text{ i.e. } \varnothing_T \,/\, \varnothing_B = 1/B$$

Thus the ratio of the cone-drum diameters is inversely proportional to the bobbin diameter. Furthermore, since we have a belt drive, the sum of the cone-drum diameters must remain constant, i.e.

$$\varnothing_T \,/\, \varnothing_B = k_2.$$

We can now use these relations to determine **K** for our example.

When the bobbin diameter, **B**, was 3.35 cm, the ratio $\varnothing_T \,/\, \varnothing_B$ was 1.845, $\varnothing_T$ was 16.86 cm, and $\varnothing_B$ was 9.14 cm. Substituting these values in Eq. (5.9), we get

$$18.45 = K/3.35$$

and hence

$$\mathbf{K} = 18.45 \times 3.35 = 6.18.$$

Thus, for any bobbin diameter, we can calculate the diameter ratio for the cone drums. We can also determine the values of the cone-drum diameters.

Consider the ratio at a bobbin diameter B, i.e.: $\varnothing_T \,/\, \varnothing_B = 618/B$ from which

$$\varnothing_B = \varnothing_T \, B/6.18$$

Substituting for $\varnothing_B$ in the equation $\varnothing_T + \varnothing_B = 26$, we obtain:

$$\varnothing_T + \varnothing_T \times (B/6.18) = 26,$$

i.e.: $\qquad \varnothing_T \, (1+ (B/6.18)) = 26,$

from which

$$\varnothing_T = 26 \times 16.8/(16.8 + B'), \varnothing_T \text{ and}: 160.68 \,/(16.8 + B'),$$

From this final equation, we can calculate directly the top-cone-drum diameter for any bobbin diameter and then obtain the bottom-cone-drum diameter by subtraction.

Example 3.1: What should be the value of the top-cone-drum diameter when the bobbin is empty, i.e., B 3.35 cm?

We have $\varnothing_T = 160.68 \,/\, (16.18 + 3.35) = 160.68/9.53 = 16.85$ cm and hence

$$\varnothing_B = 26 - 16.85 = 9.15 \text{ cm}$$

Example 3.2: When the bobbin is 10 cm in diameter, what should the cone drum diameters be?

Here we have

$$\varnothing_T = 160.68/ (6.18+10) = 160.68/ 16.18 = 9.93\text{cm}.$$

The corresponding bottom-cone-drum diameter = 26 – 993 = 1607 cm. The last value is just outside the maximum cone-drum diameter of 16 cm, so the maximum bobbin size will be a little less than 10 cm.

With the help of this constant of 160.68, a Table 3.2 can be drawn up and the top and bottom cone drum diameters listed to correspond with the bobbin size at any instant.

Table 3.2 Cone-drum dimensions in relation to bobbin diameter

Bobbin diameter, B (cm)	Top-cone diameter, = $\varnothing_T$ =160.68/(6.18+B)	Bottom-cone diameter, $\varnothing_B$ = 26 H $\varnothing_T$
3	175	85
3.35	1686	914
4	1579	1021
5	1437	1163
6	1319	1281
7	1219	1381
8	1133	1467
9	1059	1541
10	9.93	1607
11	9.35	1665
12	883	1717
13	838	1762

From the data in our example, i.e., the cone profiles are shown in Fig. 3.5. The mechanism for moving the driving belt is not shown in the gearing plan of Fig. 3.6. In some machines, the cones with hyperbolic profiles are replaced by straight-profile cone drums, and a specially designed cam is used to move the belt the required amount. The net effect is the same, namely, control of the bottom-cone-drum speed and hence control of the excess revolutions of the bobbin over the spindle in order to wind on the roving delivered from the front rollers.

Variation of the bobbin rotation speed originates in the cone transmission and occurs in small steps through shifting of the cone belt after each lift stroke. Bobbin rotation must be changed in accordance with a linear function. Unfortunately, shifting the belt by constant amounts on straight-sided cones does not vary the transmission ratio in a linear manner and thus does not produce the required linear variation in bobbin rotation speed. To obtain the desired linear variation function, the cone faces have been made hyperbolic (see Fig. 3.5), namely convex on the upper driving cone

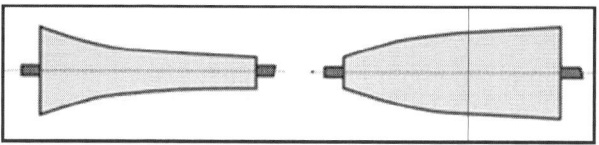

3.5 Convex and concave cones

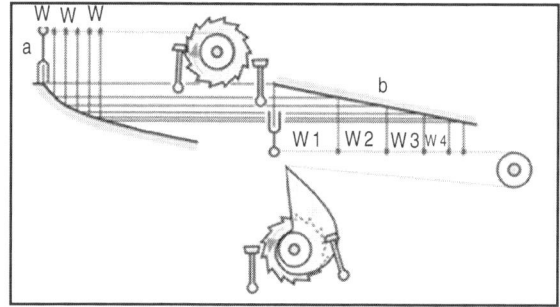

3.6 (a) Shifting the belt with hyperbolic and (b) straight-sided cones.

and concave on the lower driven cone. Hyperbolic cones are difficult to design. Additionally, during the winding operation, the belt is then always moved on surfaces of varying inclination. As a result cones are now mostly made straight-sided. However, in transmissions of this kind the belt must be shifted through steps of varying magnitude, the initial steps being relatively large (Fig. 3.6, W_1) and the later ones smaller (W_4). Instead of a hyperbolic profile on the cones (left), an eccentric is used in the belt-shifting mechanism (right).

4

Types of cams in textile and their design

4.1   Introduction

Cam is a machine component that either rotates or moves back and forth (reciprocates) to create a desired motion in a contacting element known as a follower. The shape of the contacting surface of the cam is determined by the desired motion and the profile of the follower. Cam-follower mechanisms are particularly useful when a simple motion of one part of a machine is to be converted to a more complicated desired motion of another part, one that must be accurately timed with respect to the simple motion and may include periods of dwells. Cams are essential elements in automatic machine tools, textile machinery, sewing machines, printing machines, and many others. If the follower is not restrained by a groove on the cam, a spring is necessary to keep the follower in contact with the cam. Cam systems can replace relatively complicated linkages in achieving desirable motion cycles.

4.2   Methods of driving cams

In all cam systems it is important that the follower is always in contact and following the motion of the cam. This is achieved in a number of ways including the following.

- Gravity
- Using a mechanical constraint system, i.e. groove
- Using a spring force
- Using a pneumatic or hydraulic force

Cams are made in a variety of forms, including

- A rotating disk or plate with the radial required profile
- A reciprocating wedge of the required shape
- A cylindrical barrel cam with a follower groove cut in the diameter
- A cylinder with the required profile cut in the end (end cam)

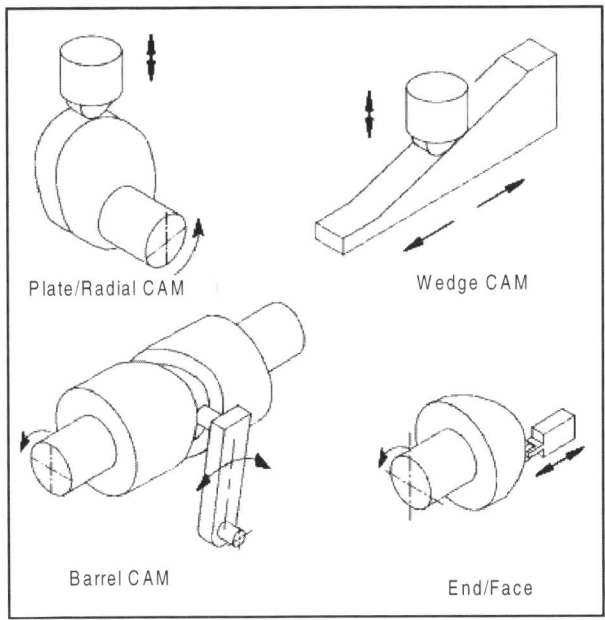

4.1 Forms of cams.

4.3   Cam followers

Cam followers can be either reciprocating or pivoting. There are various methods of transferring the motion from the cam to the follower including the following:

- Knife edge
- Flat-face
- Roller
- Curved-shoe/spherical

The cam follower can be either offset (as shown below) or in line with the cam centre of rotation.

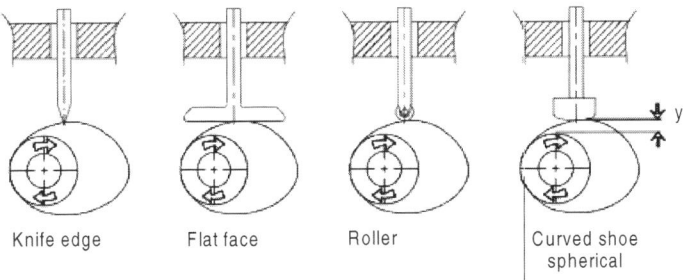

4.2 Contact methods of cam followers.

4.4   Types of cams

A cam is a reciprocating, oscillating or rotating body which imparts reciprocating or oscillating motion to a second body, called the follower, with which it is in contact. The shape of the cam depends upon its own motion, the required motion of the follower and the shape of the contact face of the follower. Of the many types of cam, a few of the most common are shown in Fig. 4.3.

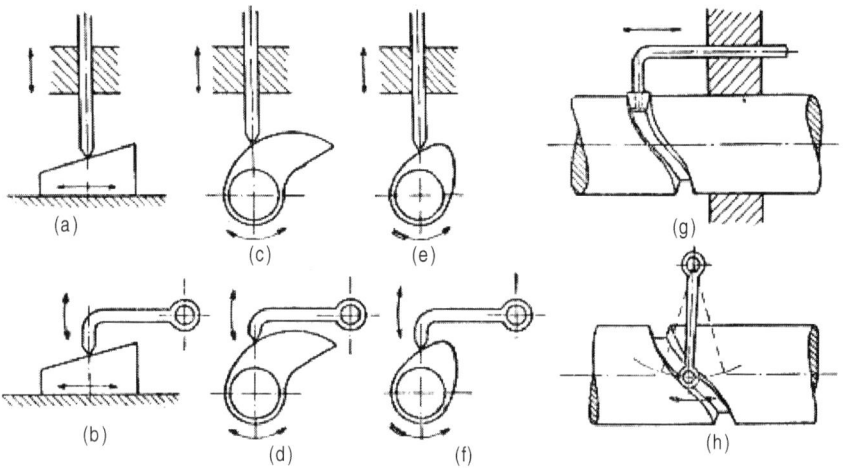

4.3 Types of cams.

In general the motion of the follower is only determined positively by the cam during a part of each stroke whilst during the remainder of the stroke contact between the cam and the follower has to be maintained by an external force, often supplied by a spring. In this connection it should be noticed that the cam does not, as would at first appear likely, to determine the motion of the follower during the whole of the its out-stroke. Actually, owing to the inertia of the follower, it is only during the first part of the out-stroke and the latter part of the return that the motion of the follower is positively controlled by the cam.

Cams are classified according to the direction of displacement of the follower with respect to the axis or oscillation of the cam. The two most important types are

- *Disc or radial cams* – In these the working surface of the cam is shaped that the reciprocation or oscillation of the follower is in a plane at right angles to the axis of the cam (see examples c, d, e, f above).
- *Cylindrical cams* – These are often used in machine tools and the

cam imparts an oscillation or reciprocation to the follower in a plane parallel to the axis of the cam (see examples g and h above).

4.5   Types of followers

Followers can be divided according to the shape of that part which is in contact with the cam. Fig. 4.4 shows some of the more common types.

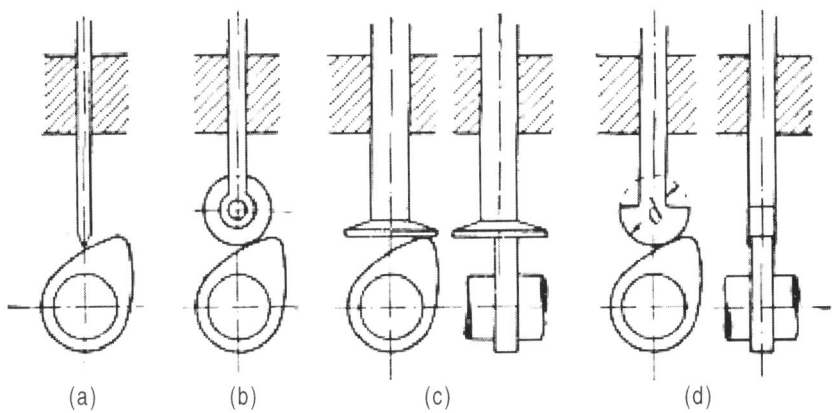

4.4 Types of followers.

- *Knife edged* (a) – These are not often used due to the rapid rate of wear of the knife edge. This design produces a considerable side thrust between the follower and the guide.
- *Roller follower* (b) – The roller follower has the advantage that the sliding motion between cam and follower is largely replaced by a rolling motion. Note that sliding is not entirely eliminated since the inertia of the roller prevents it from responding instantaneously to the change of angular velocity required by the varying peripheral speed of the cam. This type of follower also produces a considerable side thrust.
- *Flat of mushroom follower* (c) – These have the advantage that the only side thrust is that due to friction between the contact surfaces of can and follower. The relative motion is one of sliding but it may be possible to reduce this by offsetting the axis of the follower as shown in the diagram. This results in the follower revolving under the influence of the cam.
- *Flat-faced follower* – These are really an example of the mushroom follower and are used where space is limited. The most obvious example being automobile engines.

4.6   Limits imposed on the shape of the cam working surface by the choice of follower type

- The knife follower does not, theoretically, impose any limit on the shape of the cam.
- The roller follower demands that any concave portion of the working surface must have a radius at least equal to the radius of the roller.
- The flat follower requires that everywhere the surface of the cam is convex.

4.7   The cam profile for a given motion of the follower

If the required displacement of the follower is known for all angular positions of the cam, then graphical methods can be used to determine the necessary cam outline. The method of work is as follows:-

- Select the minimum cam radius i.e. zero displacement of the follower.
- Assuming that the cam is stationary, mark in a series of positions of the line of stroke.
- From knowledge of the displacements in each of these positions and allowing for the type of follower to be used, it is possible to draw the required profile of the cam (see Examples (2) and (3)).

4.8   The motion of the follower for a given cam profile

The motion imparted to the follower by a given cam profile may be determined graphically using the reverse process that was described in the last paragraph (see Example 1).

Certain standard shapes of cams which are made up of circular arcs and straight lines may be dealt with analytically. This is done by obtaining expressions for the displacement in terms of the cam angle and differentiating for the velocity and acceleration (see Examples 6 and 7).

4.9   The equivalent mechanism for a cam and follower

In many cases an equivalent mechanism using lower pairs can be substituted for a given cam and follower, possibly only over a limited range of stroke. If this is done the method of determining the velocity and acceleration which has been described in "Theory of machines, velocity and acceleration" can be used. A cam whose profile is made up of circular arcs and tangents is usually amenable to this treatment. The resulting mechanism varies with the type of follower.

When a roller follower is used, a constant distance is maintained between the centre of the roller and the centre of curvature of the cam profile. This can be replaced by a rigid link. If the follower reciprocates as in Examples 5 and 6, an equivalent slider crank chain is produced. If the follower oscillates as in the Fig. 4.5, then the motion is equivalent to a four bar chain O_1ABO_2 connecting the centers of cam axis, profile curvature, roller, and follower axis.

A flat footed oscillating follower can usually be replaced by a slotted lever (see Example 4).

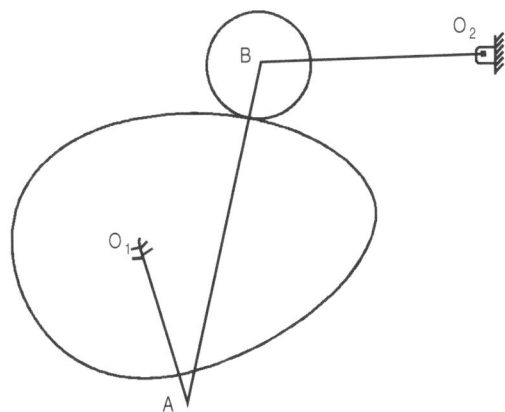

4.5 Oscillation of follower.

4.10   Negative cams and positive cams

4.10.1 Negative cams

Only one movement is given to the follower by means of cam and another movement is by means of gravity or spring.

Disc cam

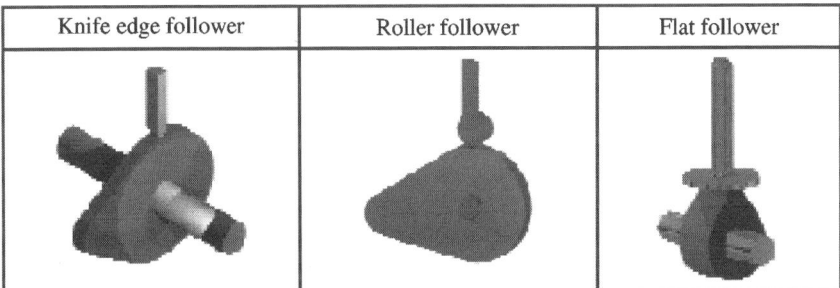

Knife edge follower	Roller follower	Flat follower

4.6 Disc cam.

Cylindrical cam

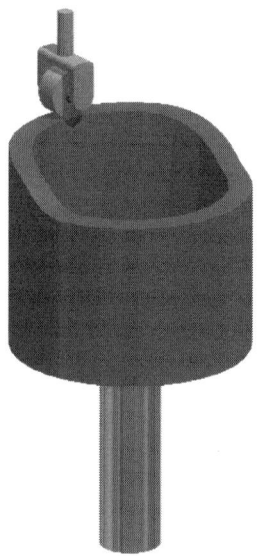

4.7 Cylindrical cam.

4.10.2 Positive cams

Both upward and downward movement is given as cam itself so it is called positive cam.

Positive cylinder cam conjugate cam

$$\omega = 16 \text{ rad/s}$$

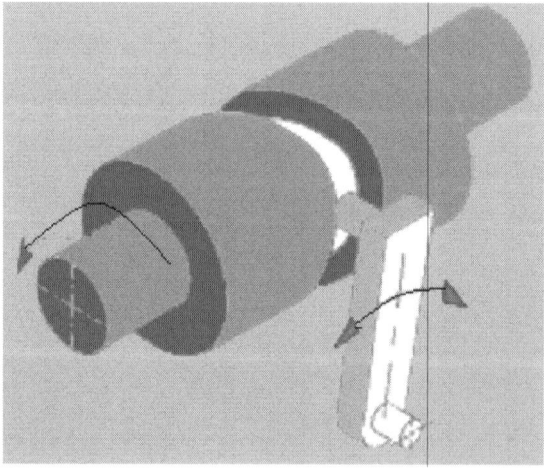

4.8 Positive Cam.

4.10.3 Textile applications

Cams are widely used in textile machines. Some good examples are shedding and picking cams in weaving, in knitting, and also in ring frame builder motion.

4.10.4 Design of shedding cams/tappets for weaving machines

To design or construct a shedding tappet the following points and dimensions must be taken into consideration:

(1) Pattern to be, produced in the fabric, that is, the number of picks in a repeat of the design and the lifting order.
(2) Lift or stroke of tappet.
(3) Distance from the centre of the driving shaft to the nearest point of contact with the treadle bowl.
(4) Time during which the healds will remain stationary, that is, the dwell of tappet
(5) Diameter of the treadle bowl

4.11   Lift of tappet

To find out the required lift of a tappet, the number of picks to a repeat at the, design and the order of lifting the healds must be first obtained. Secondly, it is necessary to ascertain the depth of the shuttle to be used, when measured along its front edge. Thirdly, other points to be measured are (a) the distance of the front edge of the shuttle from the last pick of weft in the fabric, when the shuttle enters the shed, and (b) the distance from the last pick to the healdshaft.

Lastly the depth of the shed should only be sufficient to allow the shuttle to pass; therefore, the 'lift' or 'stoke of the heald' is dependent upon the depth of the shuttle used. The shed, when opened, should remain open only long enough to allow the shuttle to pass through.

4.11.1 Lift of tappet for back healds

As the back heald is required to give a greater lift than the front heald to obtain a uniform depth of shed, the tappet operating the back heald must be constructed to give a greater lift.

4.11.2 Distance from tappet shaft to treadle bowl

The distance from the centre of the tappet shaft to the smallest part of the tappet surface varies according to position in the loom and the number of picks in one repeat. This distance for plain weave tappets placed under the healds is 1.25 in. or 3.175 cm or a little more. Tappets for 3, 4 or 5 picks to the round, similarly placed, vary from 2 to 2.5 in. or 5.08 to 6.35 cm. If placed at the loom end for 3, 4 or 5 picks, they vary from 2.5 to 3 in. or 6'35 to 7'62 cm. Above 5 picks to the round, tappets are increased considerably in size. A woodcraft tappet for 3 to 16 picks has 17 in. or 43.18 cm in diameter, and larger still for higher picks.

4.11.3 Dwell of tappet

(i)   The tappet should be so made that healds will remain stationary while the shuttle passes through the shed. This stationary period is known as "pause" or "dwell of healds", but the duration must vary to suit the fabric to be made. The length of dwell of the shedding tappet or how long the shed will remain open, will be regulated, partly as the warp threads require to be spread and partly according to the width of a loom.

(ii) The dwell of tappet must be as short as possible. Maximum time must be given for the opening and the closing of a shed. This will cause less strain and jerk upon the warp threads, and prevent their breakages. The pause or dwell of healds must be shorter for fine or tender warp, and longer in case of coarse, strong or elastic warp or wide looms.

(iii) Dwell has certain effects on the 'cover' of a cloth. In certain textures, the warp threads may run in pairs without the pieces being considered defective. They are then said to be 'reed-marked' or without cover, and the tappets employed may be made with just sufficient pause to allow the shuttle to pass through the warp; and that varies from quarter to three by eight of a pick. In 'covered cloths' all the threads are equidistant, for such and also for heavy fabrics a dwell of 1/3, 2/3 or 1/3 of a pick should be given.

(iv) The correct dwell for a tappet may be ascertained as follows:

(a)   Divide the circle described by the cranks into twelve equal parts as shown in Fig. 4.9.

(b)   In the under pick loom, 1 represents the reed in contact with the cloth and the shuttle begins to move at 4; at 5 it will enter the shed at 9; it will leave it at 10; it will be stationary in the opposite box. From 9 to 5 equals 2/3 of a revolution. This is, therefore, the maximum time to allow for changing the position of the healds; and from 5 to 9, which is 1/3 of a revolution, all the healds must be stationary. This 1/3 of a pick, that is 120° revolution of the crankshaft, is the dwell of shedding tappet.

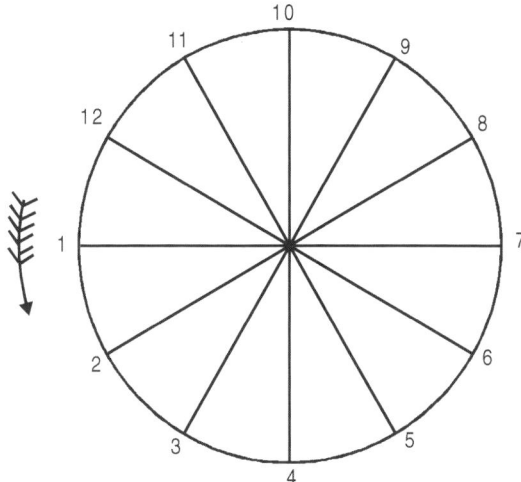

4.9 Division of crank circle.

(c)   But in certain 'covered cloths' woven in the overpick loom the dwell of tappet is from 2 to 6. In this loom the shuttle begins to move at 3; at 4 is enters the shed; at 8 it will leave it; and at 9 it will be stationary in the opposite box. From 2 to 6, all the healds must remain stationary, which is one-third of a revolution of the crankshaft; this is the dwell of tappet.

4.11.4 Construction of tappet for plain weave

To design a shedding tappet for plain weave, the following particulars have been taken into consideration:

(i)   Lift of tappet, 4 in. or 10.16 cm
(ii) Distance from the centre of the driving shaft to the nearest point of contact with the treadle bowl, 2 in. or 5.08 cm
(iii) Dwell of tappet, one-third of a pick, and
(iv) diameter of treadle bowl, 2 in. or 5.8 cm. Fig. 4.10 shows the design, the outline for the plain weave shedding tappet and the picks to the round in one repeat of the design.

At a radius of 2 in. or 5.03 cm describe the circle A (Fig. 4.10). This circle represents the distance from the centre of the driving shaft to the nearest point of contact with the treadle bowl. For plain weave tappet, the

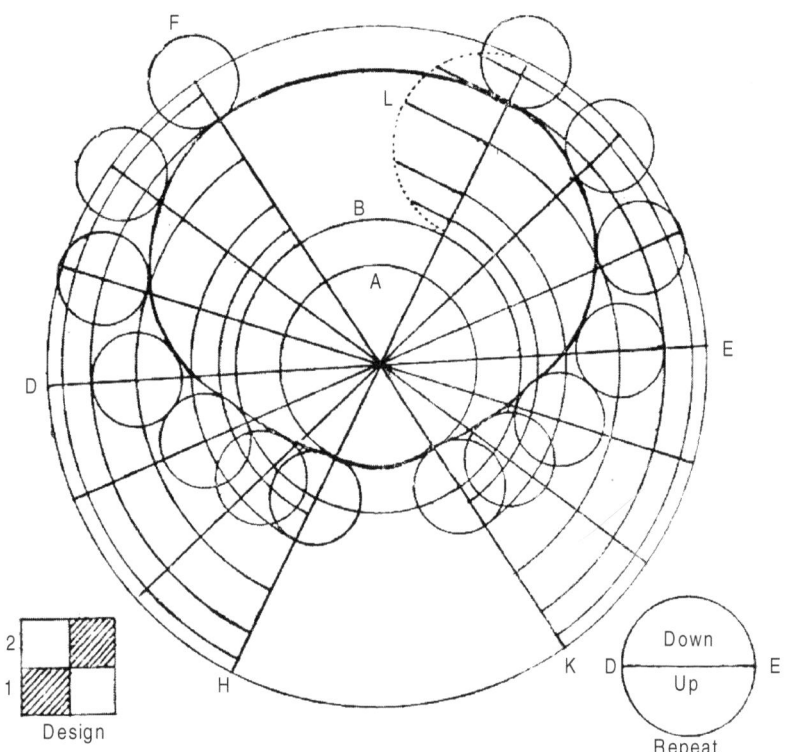

4.10 Design, tappet outline and repeat for plain weave.

bottom shaft is used as the driving or tappet shaft, whereas for twill and other weaves a countershaft is used as the tappet shaft.

At a radius of 3 in. or 7.62 cm (5.08 cm + 2·54 cm = 7.62 cm) describe the circle B. One inch or 2.54 cm is added for the radius of treadle bowl.

At a radius of 7 in. or 11.78 cm (7.62 cm +10·16 cm = 17.78 cm) describe the circle C. Four inches or 10.16 cm are added for the lift of tappet. The circle B represents the centre of the treadle bowl when the Inner circle of the tappet is acting upon the bowl. The circle C represents the centre of the howl when pressed down by the tappet.

The pattern being a plain one, as shown in Fig. 4.10, the circles must be divided into two equal parts, and each half circle will then represent one pick. By the line DE divide the circles into two equal parts.

Now as the heads must have a pause or dwell equal to one third of a pick when at the top and the bottom of their stroke, divide each half-circle into three equal parts by lines FK, GH. Divide FH and GK each into six equal parts and divide the space between the circles B and C into SIX unequal parts, the largest being in the middle, gradually decreasing towards the circles B and C. To find out the six unequal parts, describe a semi-circle L between B and C at a radius of half the lift of tappet, which is 5.08 cm. Now divide its circumference into six equal parts and then draw perpendicular lines from them on the line GH. This gives six unequal divisions on the lift to obtain the desired eccentric shape at the tappet.

From the corners of these unequal spaces, and with the radius at the treadle bowl, describe circles representing the positions of the treadle bowl at different parts of its movement. Draw the curved lines touching the extremities of the treadle bowl. This gives the outline of the plain weave tappet.

For plain weave, two tappet plates are required, which must be secured at angles of 180 degrees, the second being made fast, by the side of number one. For a three-shaft twill weave, three plates must he secured at angles of 120 degrees to each other.

For a four-shaft weave, similar parts of four plates are secured at angles of 90 degrees to each other, and so for other weaves.

Referring to Fig. 4.10, it will be seen that the treadle bowl is at rest from F to G and from K to H, or one-third of a pick at both the top and bottom of the stroke. Therefore, the time allowed for change or for moving the heald from top to bottom or vice-versa., is equal to two-thirds of a pick.If a dwell equal to half a pick is required, it can be obtained by dividing the pick into four equal parts and taking the middle two parts for dwell. If two-thirds dwell is required, divide the pick into six parts and take four parts for dwell. In plain weave looms, generally, half or one-third of a pick dwell is allowed. A dwelt of 1/3 or 1/2 of a pick is allowed for heavy and covered fabrics, but the reduction in the time for moving the healds increases the strain upon the warp yarn proportionately.

4.12   Eccentric movement of healds

As previously stated, the movement of the healds must be variable or eccentric. It must be quickest when the shed is nearly closed and must gradually decrease in speed as the shed opens. This movement spreads the warp in the fabric and causes fewer breakages. The unequal spaces into which the lift of the tappet was divided give this 'eccentric movement' to the heald. The curve of the tappet will approach nearer to the radial line as the shed closes, and the heald approaches the centre of its stroke (Fig. 4.10).

4.13   Construction of twill tappet

Draw a tappet for a 1 down, 1 up, 1 down, 1 up and 2 down (six picks to the round) twill, with the following particulars:

(i)   Centre of the tappet shaft to the nearest point of contact with the treadle bowl, 4 inches or 10.16 cm.

(ii) Lift of tapper, 2 in. or 5.08 cm.

(iii) Treadle bowl, 1.5 in. or 3.81 cm diameter and

(iv) Dwell, one-third of a pick

The method adopted for the construction of a six-shaft twill tappet is shown in Fig. 4.11.

To construct this tappet, at a radius of 4 in. or 10·16 cm, describe the circle A. At a radius of 4 in. or 12·065 cm describe the circle B. At a radius of 6.75 in. or 17.145 cm describe the circle C.

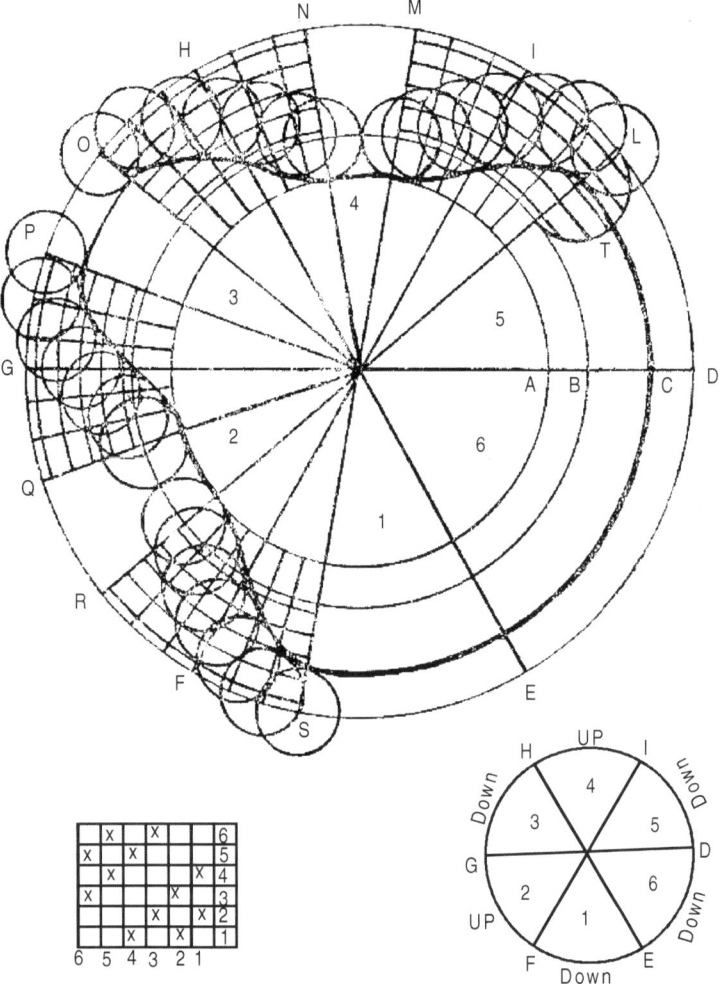

4.11 Design, tappet outline and repeat for twill weave.

As there are six picks to the round, divide the circles into six equal parts by the lines, D, E, F, G, H, I. As there is one third of a pick dwell, divide each pick into three equal parts and take the middle one for dwell. Rule the lines L, M, N, O, P, Q, R, S, to the centre, and divide the spaces allowed for change of heald into six equal parts, and the distance between the circles B and C into six unequal parts, as previously done. From the corners of the unequal spaces describe the circles representing the movement of the treadle bowl, and obtain the shape of the tappet accordingly.

It will be noticed that at the point L, the treadle bowl begins to dwell and remains stationary until it reaches the point S, when it begins to go up. The heald will, thus, be down for the first, up for the second and down for the third pick, up for the fourth, and down for fifth and sixth picks.

Eccentric movement – At a radius of half the lift of tappet, that is, 1 in. or 2.54 cm, a semi-circle T is described between the circles Band C. Its circumference is divided into six equal parts and six perpendicular lines are drawn from them. Thus the lift of tappet is divided into six unequal parts to obtain, the desired eccentric shape of the tappet to give the variable 'movement to the heald'.

For this twill, there will be six treadles, each treadle being operated by a tappet of the fame shape; but the tappet operating each succeeding treadle will be placed 60° of a revolution latter than the previous one, and so on till the sixth tappet and sixth treadle. This tappet is placed on the countershaft which must rotate once in six picks. For frequent changes of patterns, some tappets are split in 7th centre and are provided with flanged bosses to facilitate attachment to, or removal from a loom.

4.13.1 Design of builder cam

Types of cop builder

- Bobbin build
- Combined build
- Cop build

Bobbin build – Very high tension
Combined build – Chase length+ = lift-ring diameter
Cop build –

1 kg	=	9.81 N
1000 g	=	9.81 N
1 g	=	0.00981 N
	=	0.01 N
	=	1/100 N
1 g	=	1 cn (centi-newton)

Design and draw the profile of a cam.

Given data,

(1) Base diameter of cop 65 mm;
(2) nose diameter of cop = 25 mm;
(3) lift = 60 mm; (4) NPC = 25 mm;
(5) No. of leafs = 2;
(6) Winding and binding coil ratio – 1st leaf 3:1; 2nd leaf 2:5:1;
(7) Bowl diameter = 50 mm
(8) TPI = 25mm

1st leaf:

Winding angle = (360/no. of leaves) × (Winding coil ratio
(Winding coil + Binding angle coil ratio))

= [(360/2) × (3/(3 + 1))]
= 135
Binding angle = 45°

2nd leaf:

Winding angle = ((360/2) × (2.5/ (2.5 + 1)))
= 128.57
Binding angle = 51.43

S. no.	Diameter of cop	No. of revolution made of the spindle to put one coil on this diameter [πd8TPI]+1	Proportio-nate no. of revolution required by the spindle in relation to that of the highest	Proportio-nate increase required in the speed of lift (velocity)	Proportio-nate Displace-ment of bowl from procee-ding point	[Previous = value/ total*lift]	Actual Displace-ment of bowl from procee-ding point
1	65	201.98	1	1	1	0	0
2	60	186.47	0.923	1.083	1.083	4.91	4.91
3	55	171.02	0.846	1.182	1.182	5.37	10.28
4	50	155.56	0.770	1.298	1.298	5.82	16.12
5	45	140.12	0.693	1.443	1.443	6.55	22.67
6	40	124.65	0.617	1.621	1.621	7.36	30.03
7	35	109.19	0.541	1.848	1.848	8.37	38.4
8	30	93.73	0.464	2.155	2.155	9.78	48.18
9	25	78.28	0.387	2.583	2.583	11.74	59.92

Equations of motion and textile application

5.1   Introduction

This chapter deals with the basic equations of motion. Equations of motion have many applications in textile. They are used for calculation of parameters such as speed, time, etc., relating to moving parts. A number of worked out examples demonstrate their importance in the textile machinery. Fundamental definitions are given for better understanding of the concept underlying the basic equations of motion. In general, motions could be classified as linear and curvilinear. The inter-relationship between the important parameters is highlighted.

5.2   Definitions

Displacement

It is defined as the distance moved by a body from a given point.

- Types
- Linear
- Angular

Linear displacement

It is defined as the distance moved by a body in a straight line.

Angular displacement

It is defined as the distance moved by a body in an angle.

Speed

Speed is defined as the rate of change of displacement without regarding to the particular direction

Types

- Linear speed m/s
- Angular speed rad/s(ω/s)

Velocity

It is defined as the rate of change of displacement in a given direction

- Types
- Linear velocity
- Angular velocity

Relationship b/w linear and angular velocity of a (i) v = ωr body moving in a circular path is

$$v = \omega r$$

Acceleration

It is defined as the rate of change of velocity in unit time.
Its unit is m/s^2
When the velocity of the body reduces the acceleration of the body is considered to be negative and it is termed as retardation.

Equations of motion

When a body is moving at a constant rate of acceleration, the relationship between the velocities of the body at the beginning and at the end of time 't' which has passed the distance 's' is given as

(i) $v = u + at$
(ii) $s = ut + \frac{1}{2}at^2$
(iii) $v^2 = u^2 + 2as$
   where,
      u – initial velocity
      v – final velocity
      s – distance/displacement
      a – acceleration
      t – time

(i) v = u + at

$$\text{Acceleration (a)} = \text{change in velocity/time}$$
$$= (\text{final velocity} - \text{initial velocity})/\text{time}$$

$$a = (v - u)/t$$
$$at = v - u$$
$$\mathbf{u = v + at}$$

(ii) s = ut + (½ at²)

$$\text{Velocity} = \text{displacement/time}$$
$$\text{Displacement} = \text{avg. velocity} \times \text{time}$$
$$s = ((v + u)/2) \times t$$
$$s = ((u + u + at)/2) \times t$$
$$s = (2u + at)/2 \times t$$
$$s = (2ut + at^2)/2$$
$$\mathbf{s = ut + (½ at^2)}$$

(iii) v² = u² + 2as

$$v = u + at$$
$$v^2 = u^2 + a^2t^2 + 2uat$$
$$= u^2 + 2a\,[ut + ½\,at^2]$$
$$v^2 = u^2 + 2as$$

Linear	Curvilinear
$v = u = at$	$W_2 = W_1 = \alpha t$
$s = ut = 1/2at^2$	$\theta = W_1 t = 1/2\,\alpha t^2$
$v^2 = u^2 + 2as$	$W_2 = W_1^2 = 2\alpha\theta$
s = distance travelled	W_1 = initial angular velocity
t = time	W_2 = final angular velocity
a = acceleration	t = time
	α = angular acceleration
	θ = distance moved in curve path

Examples 1: A loom shuttle requires 1/12 of a second to pass through the warp sheet having 1.219 m width. Find the average velocity if it is subjected to the retardation of 9.754 m/s².

Solution: Given that t = 1/12s; s = 1.219 m; – a = 9.754 m/s².

$$s = ut + (1/2\ at^2)$$
$$1.219 = u\ (1/12) + (1/2(-9.754))\ (1/12)^2$$
$$1.219 = (u/12) - 0.11$$
$$u = 49.32 \text{ ft/s}$$
$$v = u + at$$
$$= 49.33 + (-32)\ (1/12)$$
$$v = 46.66 \text{ ft/s}$$

Average velocity = [(49.33 + 46.66)/2] = 47.99 ft/s

Example 2: A yarn guide of winding machine makes 48 doubled traverse/min., each traverse being 25 cm. Calculate the velocity of yarn guide.

Solution: Double traverses/min = 48, No. of traverse/min = 48 × 2 = 96 and Distance = 25 cm

$$\text{Velocity} = \text{Distance travelled/Time}$$
$$= 2 \times 48 \times 0.25/60$$
$$= 0.4 \text{ m/s}$$

Example 3: A machine is started from rest at full speed of 180 rpm with uniform acceleration of 36 rad/s². Find the time taken to reach full speed and angle moved by the driving shaft during acceleration.

Solution: $N_1 = 0$, $W_1 = 0$ rad/s and $N_2 = 180$ rpm

$$W_2 = [(2\pi N)/60]$$
$$= [2\pi(180)]/60 = 18.84 \text{ rad/s}$$
$$\alpha = 36 \text{ rad/s}^2$$

$$W_2 = W_1 + \alpha t$$
$$18.84 = 0 + (36)(t)$$

$t = 0.523$ s

$$\theta = W_1 t + (1/2\alpha t^2)$$
$$= 0(0.52) + (1/2(36)(0.523)^2)$$
$$\theta = 4.923 \text{ radius}$$

Example 4: A beam of warping m/c with normal warping speed of 900 mpm is stopped with uniform retardation in 12 s. What length of yarn will run over the beam during stoppage?

(i)   If beam diameter = 46 cm. What is angular retardation?
(ii) Also find out no. of revolution made during stoppage?

Solution: t = 12 s, v = 900 mpm, R = 22.5 m

$$v = r\omega \quad [r = 0.42]$$

$W = v/r$
$r = 0.225$

$$v = 0.225(\omega)$$

$$900/60 = W_1 (0.225)$$

$W_1 = 4000$ rad/min [$W_1 = 66.67$ rad/s]
$W_2 = 0$

$$W_2 = W_1 + \alpha t$$
$$0 = 66.67 + \acute{a} \ (12)$$
$$\alpha = -5.55 \ rad/s^2$$
$$\theta = W_1 t + (\tfrac{1}{2} \ \alpha t^2)$$
$$= 66.67(12) + 1/2(-5.55)(12)^2$$
$$\theta = 400.44 \ radius$$
$$\theta = 400.44/ \ (2\pi) = 63.73 \ rev$$

Length of yarn wound $= \pi \times D \times$ No. of rev
$$= \pi \times 0.45 \times 63.73$$
$$L = 90.09 \ m$$

Example 5: Carding m/c cylinder is taking 230 revolutions in coming to rest when the belt is put on to lose pulley. If running speed is 380 rpm, retardation is uniform, how long does it take to stop?

Solution: $N_2 = 0$; $N_1 = 380$ rpm; Running – rest = 230 revolutions or $\theta = 230$

$$W_1 = (2\pi n)/60$$
$$= [2\pi(360)]/60$$
$$W_1 = 39.79 \ rad/s$$
$$W_2 = 0$$

$$W_2 = W_1 + (\alpha t)$$
$$0 = 39.79 + (\alpha t)$$
$$\alpha t = -39.79$$

Displacement $= 230$ rev
$$= 230 \times 2\pi$$
$$= 1445.13 \ rad$$

$$\theta = W_1 t + 1/2 \alpha t^2$$

$$= (W_1 + \tfrac{1}{2} \ \alpha t) \ t$$
$$1445.13 = (39.79) + 1/2 \ (-39.79)t$$
$$t = 72.63 \ s$$

Example 6: A loom shuttle is moving at 30 ft/s when it leave the warp shed it has to be brought to rest in 9 in. of its movement and it stops before crank shafts moves further 60°. The loom speed is 180 picks/min. if retardation is uniform, find whether the shuttle is stopped in time.

Assume shuttle is stopped at 60° of crank revolution, then find retardation.

Solution: t = θ/6N, u = 30 ft/min, v = 0, θ = 60°, s = 9 in. = 9/12 = 0.75 ft, N = 180 rpm

$$v = u + at$$
$$0 = 30 + at$$
$$at = -30$$
$$s = ut + 1/2\ at^2$$
$$= t\ (u + 1/2at^2)$$
$$= t\ (30 + 1/2(-30))$$
$$= t\ (30 - 15)$$
$$0.75 = 15t$$
$$t = 0.05\ s \qquad \ldots (1)$$
$$at = -30$$
$$a = (-30)/0.05$$
$$a = -\ 600\ ft/s^2$$
$$t = θ/6N$$
$$= 60/6(80)$$
$$t = 0.056\ s \qquad \ldots (2)$$

(1) and (2) are same, so the shuttle is stopped in time.

Example 7: A beater having diameter of 1.2 m revolving at a speed of 900 rpm. It is required to bring down the speed of 400 rpm before the beater makes 10 revolutions. Find the time taken and also the uniform retardation req. Calculate the length of material delivered by the beater during change of speed (from 900 to 400 rpm).

Solution: θ = 10 rev × 2π = 62.33 radius, N_1= 900 rpm and N_2= 400 rpm

$$W_1 = 2πN_1/60$$
$$N_1 = 2π\ (900)/60 = 94.24\ rad/s$$
$$W_2 = 2πN_2/60$$
$$N_2 = 2π\ (400)/60 = 41.88\ rad/s$$
$$W_2 = W_1 + αt$$
$$41.88 = 94.24 + αt$$
$$αt = 52.36$$
$$θ = W_1t + ½\ αt^2$$
$$= t\ (W_1 + ½\ αt)$$
$$= t\ (94.24 + ½\ (-52.36))$$
$$62.83 = t\ (94.24 - 26.18)$$

62.83 = 68.08t

t = 0.923 s

a = –52.36/0.923

a = –56.72 rad/s²

Length of material = πDN = π (1.2) (10) = 37.69 mpm

Example 8: A beater blade of scutcher revolves at 1500 rpm in a circle of 20 cm radius. What is the angular speed in the rad/s and linear speed of the blade. If the driving pulley on the shaft of 25 cm diameter, what is the linear speed of blade?

Solution: N = 1500 rpm, r = 20 cm

w = 2πn/60 = 2π (1500)/60

w = 157.07 rad/s
v = rw
= (20/100) × 157.07
v = 31.41 m/s
d = 25 cm
v = 157.07 × (25/(2 × 100))
v = 19.63 m/s

Example 9: A shuttle is moving at 12 m/s when it enters warp shed and moves 1.6 m before it is stopped. During its passage it is subjected to uniform retardation of 12 m/s². Find time it makes to traverse the shed and its velocity when leaving.

Solution: u = 12 m/s, a = –12m/s², s =1.6 m
s = ut + ½ at²
= t (u + ½ at)
1.6 = t (12+ ½ (–12) t)
1.6 = 12t – 6t²
6t² – 12t + 1.6 = 0
t = 1.85 or 0.14 s
v = u + at
v = 12 + (–12) (0.14)
v = 10.32 m/s
t = 0.14 s

Example10: A m/c is started from rest to the full speed of 180 rpm to the uniform acceleration of 36 rad/s². Find time taken to reach the full speed and angle moved y the driving shaft during acceleration.

Solution: Given: $N_1 = 0$, $N_2 = 180$ rpm, d= 36 rad/s^2 and $W_1 = 0$

$W_2 = 2\pi \, (180)/60 = 18.84$ rad/s

$W_2 = W_1 + \alpha t$

$18.84 = 0 + (36) \, t$

$t = 0.523$ s, $\quad \theta = W_1 t + \frac{1}{2} \alpha t^2$

Example 11: A traveller in a ring frame is rotating at 8400 rpm calculate its angular speed in rad/s. if the speed is reduced to 2000 rad/s and the ring rail had to a diameter of 10 cm. Calculate the surface speed of traveller.

Solution: N = 8400 rpm

$\qquad W = 2\pi N/60 = 2\pi \, (8400)/60 = 879.64$ rad/s

$W = 879.64$ rad/s

$\qquad v = rw \quad$ [When w = 2000 rad/s]

$\qquad v = 5/100 \times 2000$

$v = 100$ m/s

Example 12: A yarn of linear density 50 tex is wound on a empty package continuously for 20 min. At the end of this time the mass of yarn on the package is found to be 47.1 g. Calculate the winding speed.

Solution:

50 tex [tex = 1 g in 1000 m]

50 tex = 50 g in 1000 m.

Weight of yarn on package wound for 20 min is = 1000/50 × 47.1

$\qquad\qquad\qquad\qquad\qquad\qquad\qquad\qquad = 942$ m for 20 min.

$\qquad$ (i.e.) the length wound per minute = 942/20 = 47.1 mpm

Example 13: Carding engine was absorbed to take 24s to get up to the full speed of 160 rpm. What is the acceleration and how many revolutions are made during the acceleration.

Solution: T = 24 s, $N_2 = 160$ rpm

$\qquad\qquad W_2 = 2\pi \, (160)/60 = 16.75$ rad/s

$\qquad\qquad W_1 = 0$

$\qquad\qquad W_2 = W_1 + \alpha t$

$\qquad\qquad 16.75 = 0 + 24\alpha$

$\qquad\qquad \alpha = 0.697$ rad/s^2

$\qquad\qquad \theta = W_1 t + \frac{1}{2} \alpha t^2$

$\qquad\qquad\quad = 0 + \frac{1}{2} \, (0.698) \, (24)^2$

$\qquad\qquad \theta = 201.02$ rad

No. of revolutions = $\theta/2\pi = 201.2/2\pi = 31.99$ revolution

6

Energy, force and moments in textile machines

6.1   Introduction

This chapter deals with the main types of energies, namely kinetic and potential energies. These types of energies are useful in textile applications. Various moving parts of textile machinery have these energies. Some numerical examples are given to illustrate as to how these energies could be calculated with the relevant formulae.

Energy

Energy is defined as the capacity of doing work. It may be existing in different forms such as potential, kinetic, thermal, pressure, strain, electrical, electromagnetic and chemical energy.

Energy may be stored but the method of storage depends on the type of energy involved.

Potential energy

The energy possessed by the body by the virtue of its position.

It depends on the position of the body above datum reference line, is stored by retaining the body in that position.

Its value depends on the weight of the body and the height above the reference level it is been placed.

Obviously this value is equivalent to the work done in elevating the body to its position from the reference level.

And this amount of energy is stored by preventing the return of the body back to the reference level as long as movement of the body is against gravity is restrained.

Kinetic energy

It may be defined as the energy possessed by the body by virtue of its motion.

It obviously is not stored but the concept of storage applied to electromagnetic energy is somewhat nebulous.

Common unit of energy is naturally same as that of work.

Its unit is joule

$$1 \text{ joule } = 1 \text{ Nm}$$
$$1 \text{ watt } = 1 \text{ Nm/s or } 1 \text{ joule/s}$$

Work and energy

If an object is moved as long as a distance of 1 m by a force of in then the work done is 1 joule.

This is known that it is not possible to recover the amount of work done and if we wish to return the object to its original position, we must do the further 1 joule of work for the purpose.

On the other hand if we lift an object to 1m from the reference level by a force of 1N then the work done of 1 joule can be recovered reasonably completely if the object is released and allowed to fall back to its original position as long as some mechanism for making use of the work is available.

In the first case, the work is being done against friction and in latter the work is being done against gravity.

In the first case, the work done against friction is converted into heat and the heat energy produced is dissipated and not easily recoverable.

But in second case, it is possible for the force of gravity to perform some work and the potential energy stored can be recovered by nearly releasing the lifted object.

6.2   Deriving the expressions for potential and kinetic energy

Kinetic energy

An object of mass in moving with velocity v is brought to rest in a distance by the resisting action of a constant force f.

$$KE = \text{ work done}$$

$$KE = F \times S \qquad \qquad \dots (1)$$

We know,

$$F \quad = m \times a \qquad \qquad [\therefore \text{the object is brought to rest, a is considered as retardation}]$$

$$\therefore \qquad KE = mas \qquad \qquad \dots (2)$$

By considering motions

Initial velocity = u
Final velocity = $\vartheta = 0$
A = – a

Using

$$\vartheta^2 = u^2 + 2as$$
$$0 = u^2 - 2as$$
$$u^2 = +2as$$

$$\frac{u^2}{2} = as \qquad\qquad \ldots (3)$$

Apply eqation (3) in Q

$$KE = \frac{U^2}{2} m$$

$$\mathbf{KE = \tfrac{1}{2}\, mu^2}$$

Expression for potential energy

The energy released by the fall of an object of mass 'm' through a height 'h' to the ground is equivalent to the work done during fall.

$$PE = \text{work done}$$
$$PE = \text{Force} \times \text{distance moved}$$
$$\mathbf{PE = mgh} \quad [g = \text{gravitational acceleration}$$
$$= 9.81$$

6.3   Worked examples

Example 1: A bale of wt 100 kg comes down through a chute as shown in fig. The kinetic energy and the potential energy is to be found out at the top and bottom of the chute and also at floor level. Also calculate the time taken for traverse, the total time of fault and the horizontal distance moved by the bale.

Solution:

$$KE_1 = 0$$
$$PE_1 = mgh$$

$$= \frac{w}{g} \times g \times h$$

$$= wh$$
$$= (100)\,(25)$$
$$\textbf{PE}_1 = \textbf{2500 kgm}$$

$$PE_2 = mgh$$
$$= 100 \times 16$$
$$PE_2 = 1600\ kgm$$

$$KE_2 = PE_2 - PE_2$$
$$[\therefore \text{Sum of potential energy}$$
$$\text{and kinetic energy will be equal}]$$
$$= 2500 - 1600$$
$$KE_2 = 900\ kgm$$

$$\textbf{KE}_2 = \frac{\textbf{1}}{\textbf{2}}\textbf{mv}^2$$

$$900 = \frac{1}{2} \times \frac{100}{9.81} \times v^2$$
$$V = 13.28\ m/s$$

$$\sin 60° = \frac{9}{S}$$

$$S = \frac{9}{\sin 60°}$$
$$\textbf{S = 10.39 m}$$

$$v^2 = u^2 + 2as$$
$$13.28^2 = 0 + 2a\,(10.39)$$
$$\textbf{a = 8.48 m/s}^2$$

$$v = u + at$$
$$13.28 = 0 + 8.48\ t$$
$$\textbf{t}_1 = \textbf{1.56 s}$$

$$\text{Horizontal component} = v \cos 60°$$
$$= 6.64\ m/s$$
$$\text{Vertical component} = v \sin 60°$$
$$= 13.28\,(\sin 60°)$$
$$\vartheta = 11.50 m/s$$

$$v^2 = u^2 + 2as$$
$$= 11.5^2 + (2 \times 9.81 \times 16)$$
$$\mathbf{v = 21.12 \ m/s}$$

$$v = u + at_2$$
$$21.12 = 11.5 + 9.81 \ t_2$$
$$t_2 = 0.98 \ s$$
$$t = t_1 + t_2$$
$$= 1.56 + 0.98$$
$$= 2.54 \ s$$

$$\text{Horizontal landing distance} = v \cos\theta \ (t_2)$$
$$= V \ 3.28 \cos 60° \ (0.98)$$
$$= 6.50 \ m$$
$$PE_3 = 0$$
$$KE_3 = 2500 \ kgm$$
$$[\therefore KE + PE \text{ are equal at 3 cases}]$$

$$2500 = \frac{1}{2}mv^2$$

$$V = \frac{2500 \times 2 \times 9.81}{100} \qquad \left[\therefore m = \frac{w}{g} \right]$$
$$\mathbf{v = 22.14 \ m/s}$$

Potential energy cannot be destroyed

$$\text{Work done} = KE$$

$$\theta = \tan^{-1}\left[\frac{v}{h}\right] = \left[\frac{22 - 14}{13.28 \cos 60°}\right]$$
$$\theta = 73°$$

Example 2: A carding engine cylinder revolving at 180 rpm has 35,000 joules of kinetic energy stored in it when the belt is shifted of the m/c comes to rest in 130 revolutions. Assuming that the KE is the total for all the moving parts of the m/c, find the power requirement to run it at its operating speed of 180rpm and also the time requirement to stop the m/c.

Solution: Given,
$$N_1 = 180 \ rpm$$
$$KE = 35,000 \ joules$$
$$\text{No. of rev. taken by the m/c to stop } = 130 \ rev.$$
$$= 130 \times 2\pi \ rad$$
$$= 816.81 \ rad$$

$$N_2 = 0$$

$$W_1 = \frac{2\pi N}{60} = \frac{2\pi(180)}{60}$$
$$w_1 = 18.84 \text{ rad/s}$$

$$w_2 = 0$$

$$w_2 = w_1 + \alpha t$$
$$0 = 18.84 + \alpha t$$
$$\alpha t = -18.84$$

$$\theta = w_1 t + \frac{1}{2}\alpha t^2$$
$$816.81 = 18.84t + 1/2\ (-18.84)t$$
$$= 18.84t - 9.42t$$
$$\mathbf{T = 86.71\ s}$$

For 130 rev, KE stored is 35,000 joules $= \dfrac{35,000}{130}$

$$= 269.23 \text{ joules}$$

For 1 s, no. of rev $\qquad = \dfrac{180}{60} = 3 \text{ rps}$

For 1 s, no. of joules $\qquad = 269.23 \times 3$
$$= 807.69 \text{ joules}$$
(watts)

Example 3: A card cylinder is disconnected from its drive and from the remaining parts of the m/c so that it can rotate freely in its bearings for a test as bearing friction. In this best a rope is coiled around a pulley on the cylinder shaft passed around a frictionless pulley system suspended from the roof and connected to 20 kg weight this wt is allowed to fall from the height of 3 m to the ground. It takes is in doing so and rotates the cylinder through 1,5 rev. At the instant that it strike the ground, the rope disengages itself from the pulley and the cylinder comes to rest after further 7 rev. Calculate KE in weight when it reaches the floor, in the cylinder at the same instant, and in the cylinder running at its normal speed of 3 rps on the assumption that the frictional resistance at the bearings remains unchanged throughout the various arrangements.

Solution: Given: u = 0, t = 9s, s = 3 m

Considering the motion f falling weight

$$s = ut + \frac{1}{2}at^2$$

$$3 = 0 + \frac{1}{2}a(9)^2$$

$$a = 0.074 \text{ m/s}^2$$

W.K.T

$$v = u + at$$

$$= 0 + 0.074 \text{ (a)}$$

$$v = 0.666 \text{ m/s}$$

KE of a falling weight as it reaches the floor is

$$KE = \frac{1}{2}mv^2$$

$$= \frac{1}{2}\left(\frac{20}{9.81}\right) \times 0.666^2$$

$$KE = 0.45J$$

Potential energy of a falling weight as it reaches the floor is

$$= mgh$$

$$= \frac{20}{9.81} \times 9.81 \times 3$$

$$P.E. = 60 \text{ Nm or J}$$

Energy transferred to the cylinder = Original P.E – K.E. in ut as it
reaches the floor

$$= 60 - 0.44$$

$$= 59.56 \text{ J}$$

Energy absorbed by the friction system per revolution

$$= \frac{\text{Actual Energy Transferred}}{\text{Total No. of Revolutions}}$$

$$= \frac{59.56}{(7+1.5)}$$

$$= 7.01 \text{ J}$$

Total energy absorbed $= 7.01 \times 1.5$

$$= 10.51 \text{ J}$$

Actual energy transferred to cylinder $= 59.56 - 10.515$

$$= 49.045 \text{ J}$$

Consider u = 0, t = 9s, S = 1.5 rev

$$S = ut + 1/2\ at^2$$

$$1.5 = 0 + 0 + \frac{1}{2}a(9)^2$$

$$A = 0.037\ rev/s^2$$

$$V = u + at$$

V = 0.333 revolutions

Clockwise moments are also called as negative moment.
Anticlockwise moments are also called positive moments.

6.4   Principle of moment

- Principle of moment can otherwise be called as moment's condition. It can be stated as "For a body in equilibrium under the influence of any number of forces the algebraic sum of the moments of all the forces about any point in the plane will be zero".

- It means that the sum of clockwise or negative moments is exactly equal to the sum of anticlockwise or positive moments.

A simple lever arrangement for applying pressure to the rollers of a padder is used at each side of padder to achieve even pressure if the mass of the roller is 25 kg, the mass of wading bob is 40 kg, the pressure point is 15 cm from fulcrum and centre of gravity of loading bob is 60 cm from fulcrum. Calculate the force exerted on the fabric passing b/w the rollers and the force exerted by the lever on the fulcrum.

Since three forces are acting on the lever are in equilibrium, there must be no net force in any direction.

F is acting vertically downwards on the fulcrum.

Therefore, the upward pressure p and downward force f must be exactly balanced by the fulcrum.

Therefore the force, acting on the fulcrum is

$$F = 160 - 40$$

$$F = 120\ kg$$

Exercises

1. A weighing system is used for applying pressure to a pair of calendar roller. It consists of lower and upper lever connected by a vertical rod at each side of machine. Each lever has the mass of 10 kg, has its centre of gravity at 45 cm from the fulcrum and carries the load of 30 kg, at distance in from the fulcrum. The vertical rod is attached

7.5 cm from fulcrum of the lower lever and to one end of the top lever, 20 cm from its fulcrum at other end. The top lever rests on the bush of the top roller. Centre of the bush being 6 cm from the fulcrum. The top lever has a mass of 5 kg, both levers are horizontal and the top roller has a mass of 90 kg. Circular the force applied by the top roller to the material passing between them.

2. The heald shaft of the loom is attached to a horizontal lever, at a distance of 40 cm, from the fulcrum reversing spring is attached to the lever at a distance of 10 cm. From fulcrum, the lever must exert the minimum resistance of 40 N to the lifting of heald shafts. What pull must the spring exacts when the lever is horizontal?

The revolving at centre point gets all load (4kg) and roving begetting no load in asymmetric system.

In double boss system, the sum of the loads acting on roving A and B will be equivalent to the total load applied.

7
Friction in textile machines

7.1 Introduction

Friction is a force which acts in opposite direction during movement of a body over another one. When one surface has contact with another surface the ridges and depressions interlocked and thus relative motion of one surface over the other is resisted. There are two types of friction, namely, static and kinetic. It has important applications in textiles, such as in yarn passing through guides, and during passage of textile materials over machine parts. Lower friction causes slippage, which higher friction causes abrasion or breakage of material. It is therefore necessary to maintain the necessary value of friction in a number of applications.

This resistive force is known as force of friction (Fig. 7.1).

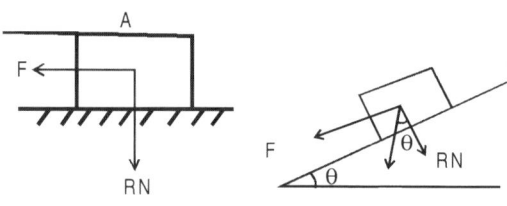

7.1 Friction.

7.2 Coefficient of friction

The ratio between limiting force to the normal reaction

$$m = \tan\theta = \frac{F}{R_N}$$

θ = angle of friction.

7.3 Static friction

It is friction of a body which is just above to start sliding over other.

7.3.1   Laws of static friction

1. The force of friction always acts in a direction opposite to that in which the body tends to move, if the force of friction would have been absent.
2. Magnitude of force of friction is exactly equal to the force which tends to move the body.
3. Magnitude of limiting friction bears a constant ratio to normal reaction between two surfaces.

   i.e., $\mu = \dfrac{F}{R_N}$

4. Force of friction is independent of area of contact between two surfaces.
5. Force of friction depends upon the roughness in the surface.

7.4   Dynamic or kinetic friction

Friction when the body is sliding at a slow uniform speed. Types of kinetic friction are

(a) Rolling friction
(b) Sliding friction
(c) Pivot friction

7.5   Friction b/w two surfaces depends upon

1. the materials of which the surfaces are composed
2. the state of surface (smoothness or roughness)
3. pressure between the surfaces

7.6   Coil friction

The friction b/w the flexible band and curved surface, i.e. coil friction (Fig. 7.2).

The difference between tight side and slack side tension is known as effective tension (Fig. 7.3).

$$T = T_t - T_s$$

ΔT = Difference between running stationery tension on either side.

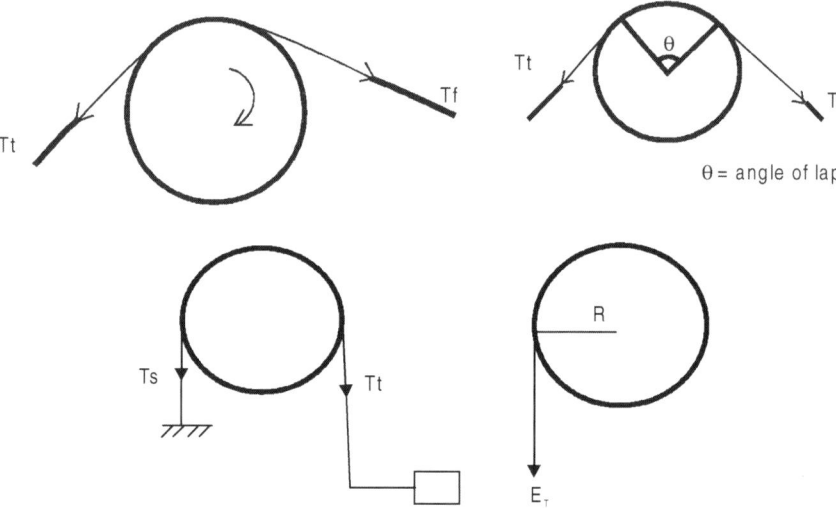

θ = angle of lap

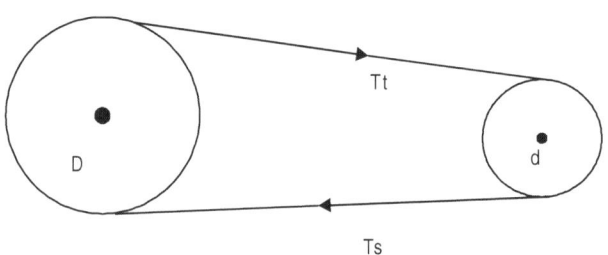

7.2 Coil friction.

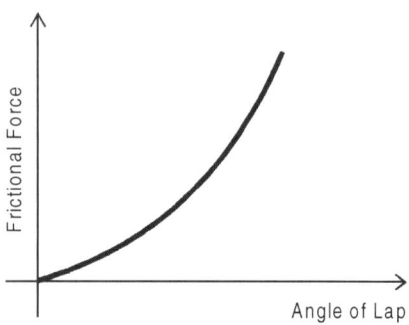

Tt = Tight side; Ts = Slack side
D = Driven pulley; d = Driving pulley

7.3 Tight side and slack side tension.

7.4 Relationship between angle of lap and frictional force.

Worked examples

Example 1: A let-off for a 100 m beam is arranged as shown in Fig. 7.5. A band fixed to the frame of the 100 m is coiled around the weighting ruffle and attached to the weighting lever which is pulled down by the weight; the ruffle diameter is 15 cm and the effect of weight is equivalent to 100 kg load suspended from the band the coefficient of friction between band rubble is 0.18. Calculate the work done against the friction for 1 rev of the beam is identical break arrangements may be assumed at each end of it.

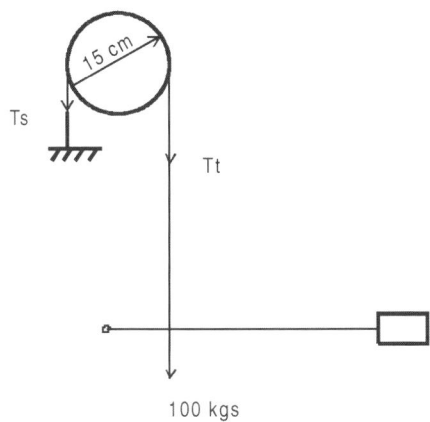

100 kgs

7.5 Let-off.

Solution:

$$\theta = 180° = \pi$$
$$d = 15\text{cm}$$
$$\mu = 0.18$$

$$\frac{T_t}{T_s} = e^{(0.18 x \pi)}$$

$$\frac{T_t}{T_s} = 1.76$$

$$T_s = \frac{100}{1.76} = 56.8\,\text{kg}$$

$$T_t = 100\ \text{kg}$$

$$F = T_t - T_s$$
$$= 100 - 56.8$$

$$= 43.2 \text{ kg}$$
$$\text{Total force} = 2 \times 43.2$$
$$= 86.4 \text{ kg}$$
$$= 847.58 \text{ N}$$

Distance moved by the ruffle/rev $= \pi \times d$

$$= \pi \times \frac{15}{100}$$
$$= 0.47 \text{ m}$$

$$\text{Work done/rev} = 847.58 \times 0.47$$
$$= 398.36 \text{ J}$$

The warp yarn is passed over a guide rod and making an angle of lap of 90°. If the between guide rod and yarn is 0.15, the input tension is measured as 10 g. Find out the output tension is measured as 10 g. Find out the output tension and also effective (When the angel of lap is 150°, 180°, 120°).

$\theta = 90° = 1.57 \text{rad}; \mu = 0.15; T_s = 10 \text{ g}$

$$\frac{T_t}{T_s} = e^{\mu\theta}$$

$$= e^{(0.15 \times 1.57)} = 1.265$$
$$T_t = (10)(1.265)$$
$$T_t = 12.65 \text{ g}$$

$$T = T_t - T_s$$
$$= 12.65 - 10$$
$$T = 2.65 \text{g}$$

$\theta = 150° = 2.61 \ rad;$

$$\frac{T_t}{T_s} = e^{\mu\vartheta} = e^{(0.15 \times 2.61)}$$

$$T_t = 10 x e^{0.3915}$$
$$\mathbf{T_t = 14.79 \text{ g}}$$

$$T = T_t - T_s$$
$$= 14.79 - 10$$
$$\boldsymbol{T = 13.79 \text{ g}}$$

$\theta = 180° = \pi$

$$T_t = 10\,e^{(0.15 \times \pi)}$$
$$T_t = 16.01 \text{ g}$$

$$T = 16.01 - 10$$
$$T = 15.01\text{g}$$

$\theta = 120° = 2.09$

$$T_t = 10\,e^{(0.15 \times 2.09)}$$
$$T = 13.68 - 10$$

Problems 2: A ring frame traveler is 0.5 g is rotating at a speed of 8000 rpm. The ring diameter is 40 mm (Fig. 7.6). Find out the power required to overcome the friction if μ is assumed to be 0.2. The total number of spindles in the ring frame is 504.

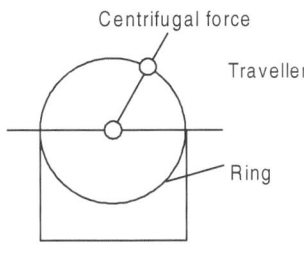

7.6 Ring and traveler.

Solution: Given: no. of spindles = 504; W = 0.5 g; θ= 40 mm; μ = 0.2;

$$\mu = \frac{0.5}{9.81} = 0.051 \text{ g} = 0.000051 \text{ kg}$$

$$\text{Centrifugal force} = \frac{mv^2}{r}$$

$$v = \pi DN = \pi \left(\frac{40}{1000}\right) \times \frac{8000}{60}$$

$$V = 16.75 \text{ m/s}$$

$$\text{Centrifugal force} = \frac{(0.051)(16.75)^2}{1000 \times \left(\frac{20}{1000}\right)}$$

$$= 0.715 \text{ kg m/s}$$

$$\mu = \frac{F}{R_N}$$
$$F = 0.2 \times 0.715$$
$$F = 0.143$$

$$\text{Power} = \text{Force} \times \text{velocity}$$
$$= 0.143 \times 16.75$$
$$= 2.395$$

No. of spindles = 504

$$\text{Power} = 2.395 \times 504$$
$$= 1207.08 \text{ watts}$$
$$= 1207 \text{ watts}$$

Problem 3: A band brake (Fig. 7.7) has diameter of the drum as 600 mm, and the band makes an angle of 240° with the drum and μ is 0.2. The drum rotates at the speed of 200 rpm. The rope is connected to the weighing lever at a point which is 5 cm from and fulcrum and the weight of 30 kg is mounted at a distance of 50 cm from the fulcrum. Calculate the work done in rev. power dissipated and the forage when the drum is moving at (a) clockwise direction and (b) anticlockwise direction.

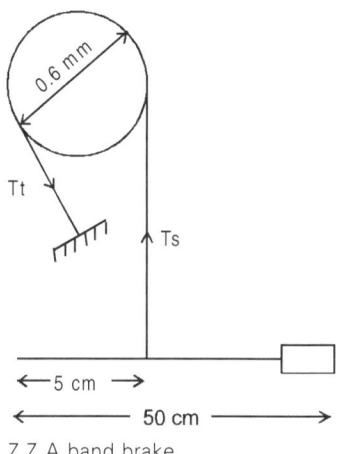

7.7 A band brake.

Clockwise direction: Taking moment about fulcrum

$$T_s \times 5 = 30 \times 50$$
$$\mathbf{T_s = 300 \text{ kg}}$$

$$\frac{T_t}{T_s} = e^{\mu\theta}$$

$$= 300 \times e^{(0.2 \times 4.18)}$$

$$T_t = \textbf{693.52kgs}$$

$$F = T_t - T_s$$
$$= 693.52 - 300$$
$$\textbf{F = 393.52 kg}$$

Work done/rev $= F \times \pi D$
$$= 393.52 \times \pi \times 0.6$$
Work done/rev = 7276.77 Nm or J

$$\text{Power} = 393.52 \times \pi \times 0.6 \times \frac{200}{60}$$

$$= 2472/55 \text{ kg m/s}$$
Power = 24.72 kw

Torque $T = (T_t - T_s)r$
$$= (393.52) \ 0.3$$
Torque T = 118.06 kg m

Anticlockwise direction, Taking moment about fulcrum

$$T_t \times 5 = 30 \times 50$$
$$\textbf{\textit{T}}_t \textbf{ = 300 kg}$$

$$\frac{T_t}{T_s} = e^{\mu\theta}$$

$$T_s = \frac{300}{e^{(0.2 x 4.12)}}$$

$$T_s = 130.03 \text{ kg}$$

$$F = T_t - T_s = 300 - 130.03$$
F = 169.96 kg

$$\text{Work done / rev} = F \times \pi D$$
$$= 169.96 \times \pi \times 0.6$$
Work done /rev = 320.38 Nm or J

$$\text{Power} = 169.96 \times \pi \times 0.6 \times \frac{200}{60}$$

$$= 1067.89 \text{ kg m/s}$$
Power = 10.67 kw

$$\text{Torque } T = (T_t - T_s)\, n$$
$$= (300 - 130.03)\, 0.3$$
$$= (169.96)\, 0.3$$
T = 50.98 kg m

Negative or friction let-off

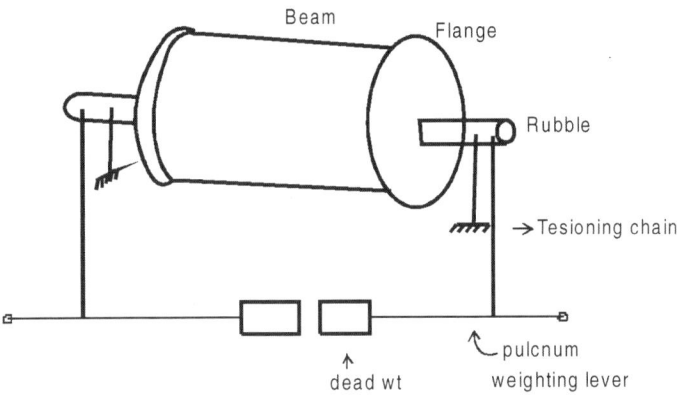

7.8 Negative or friction let-off

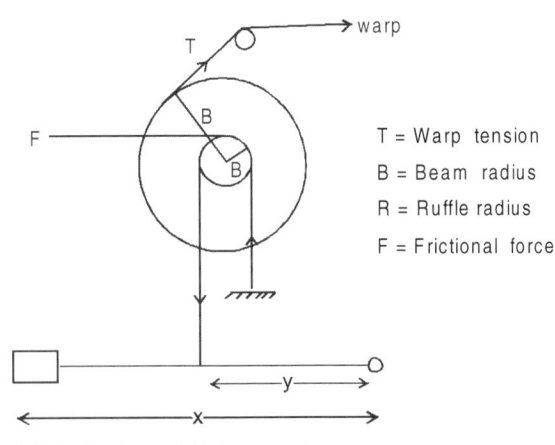

T = Warp tension
B = Beam radius
R = Ruffle radius
F = Frictional force

7.9 A simple weight lever system

The above shown system has slippage

$$T \times B = F \times R$$

$$T = \frac{FR}{B}$$

$$T\alpha\frac{F}{B} \quad (\Theta \ R = \text{constant})$$

To maintain warp tension T- constant

$$F\alpha B \qquad \qquad \dots (1)$$

We know that

$$F = T_t - T_s \qquad \qquad \dots (2)$$

$$\frac{T_t}{T_s} = e^{\mu\theta}$$

$$T_s = K_1.T \ \left(\because \frac{1}{e^{\mu\theta}} = constant \right) \quad \dots (3)$$

From the Figure 7.9

$$T_t \times Y = W \times X$$

$$T_t = W.\frac{x}{y}$$

$$T_t = K_2.x \qquad (w/y = k_2) \qquad \dots (4)$$

Applying (3) in (2)

$$F = T_t - \frac{T_t}{e^{\mu\theta}}$$

$$= T_t \left(1 - \frac{1}{e^{\mu\theta}} \right)$$

$$F = T_t.K_3 \qquad \qquad \dots (5)$$

Apply (4) in (5)

$$F = K_2.X.\ K_3$$

$$F = K_4.X$$

$$F \propto X$$

Therefore the frictional force is directly proportional to the distance of weight from fulcrum.

If the distance decreases, the fictional force 'F' also decreases.

From equation (1) it is clear that frictional force decreases, if the beam diameter decreases.

As the beam diameter is decreasing, we have to reduce the length (x) proportionally.

Problem 2: A simple weight lever system (Fig. 7.10) at each side of loom beam is provided with weight of 250 new tons. The leverage ratio (x/y) = 4. Full beam radius = 20 cm. Ruffle radius is 7 cm. No. of lap (turns) 1½ turns μ = 0.15.

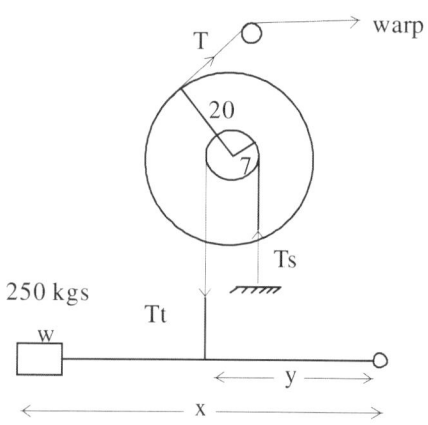

7.10 A simple weight lever system.

Find the warp tension at the point of slippage.
When it is not an angle $\theta = 180°$

$$T_t\, xy = wxx$$
$$T_t = wx/y$$
$$= 250 \times 4$$
$$T_t = 1000N$$

$$T = \frac{FR}{B}$$

$$F = \frac{(20)/(250)}{T}$$

$$T_t \text{ Angle of lap} = 540 \times \frac{\pi}{180}$$

$$\theta = 9.42$$

$$\frac{T_t}{T_s} = e^{\mu\theta}$$

$$\frac{T_t}{T_s} = e^{(0.15 \times 9.42)}$$

$T_t = 4.108\ T_s$
$T_s = 1000/4.108 = 243.4N$

$F = 1000 - 243.4$
F = 756.57 N
$F = 2 \times 756.57 = 1513.14N$
F = 1513.14 N

$T = 256.8N$
T = 2 × 756.57 × 7/20   X/2
T = 529.7 N

Problem 3: Figure 7.11 shows friction let-off weighing motion for 100 m. A rope is fixed at A to the frame and coiled around the ruffle and attached to weighing lever at B. $\mu = 0.16$ what warp tension will cause slip and let off the 1½ turns in another case. Assume similar arrangement at each side of beam.

$Y = 4"$ $x = 20"$ $w = 50$ pounds. Ruffle radius 6". Beam radius = 16"

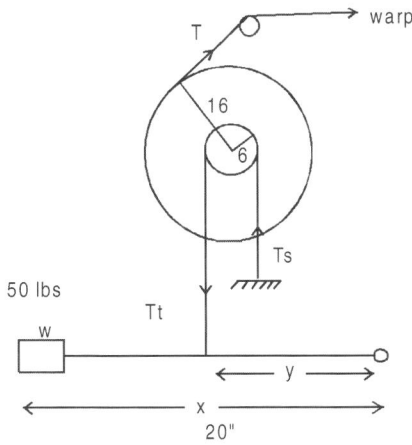

7.11 A simple weight lever system

Moment about fulcrum

$$T_t (4) = 50\ (20)$$
$$T_t = \textbf{250 lbs}$$

(i) $\theta = 180° = 3.14$

$$\frac{T_t}{T_s} = e^{mq}$$

$$\frac{250}{T_s} = e^{(0.16 \times 3.14)}$$

$$T_s = 151.26 \text{ lbs}$$

$$F = T_t - T_s$$
$$= 250 - 151.26$$
$$F = 98.7 \text{ lbs}$$

$$F = 98.7 - 2$$
$$F = 197.46 \text{ lbs}$$

$$T = 197.46 \times \frac{6}{16}$$
$$F = 74.04 \text{ lbs}$$

$\theta = 1^{1/2} \text{ turns} = 9.42$

$$\frac{T_t}{T_s} = e^{\mu\theta}$$

$$\frac{250}{T_s} = e^{(0.16 \times 9.42)}$$

$$T_t = 55.38 \text{ lbs}$$

$$F = T_t - T_s$$
$$= 250 - 55.38$$
$$F = 389.23 \text{ lbs}$$

$$T = 389.23 \times \frac{6}{16}$$
$$F = 145.96 \text{ lbs}$$

Problem 4: In figure shown the band is fixed to the loom frame and coiled the ruffle and attached to the weight equivalent to 100 kg load ruffle diameter = 15 cm, $\mu = 0.18$. Calculate work done and frictional force for 1 rev. of the beam if individual breaking system is provided at both ends.

μ = 0.18
F = ?
θ = 1 rev = π
W = 100 kg

Problem 5: A leather brake bar presses against rim of the pulley with a normal force of 120 N. If μ = 0.15. What power is dissipated in stopping the machine in 10s if the diameter of pulley is 28 cm and it complete 18 rev. in stopping process?

$$F = \mu \times R_F$$

$$= 0.5 \times 120$$

$$F = 60 \text{ N}$$

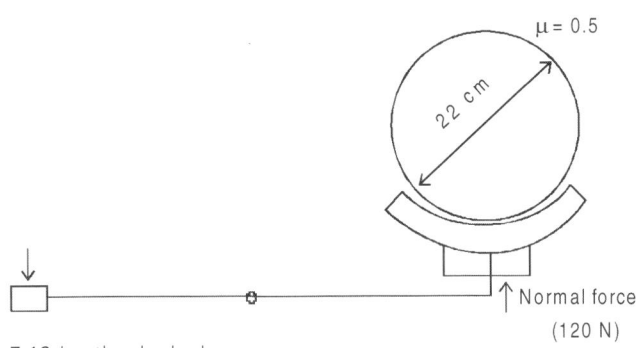

7.12 Leather brake bar.

Power = Force × Velocity

18 rev. in 10 s

For 1 s = $\dfrac{18}{10}$ = 1.8 rps

Velocity = πDN

$$= \pi (1.8) \left(\frac{28}{100} \right)$$

$$= 1.58 \text{ m/s}$$

Power = 1.58 × 60

P = 95 watts

8

Clutches and brakes

8.1 Introduction

In a machine it may be necessary to control the motion of certain element(s) depending on the need. In simpler terms, motion to any element has to be transferred or motion of that element has to be stopped depending on the situation. Clutches and brakes are used to control the flow of mechanical power. Both have similar construction and work on the same principles. However, there is a functional difference between the two. Clutch connects one moving part to another part, whereas the break connects one moving part to another stationary part. In both the cases, frictional, mechanical, electromagnetic or hydraulic or pneumatic connections exist between two elements. Brakes and clutches are extensively used in textile production. Clutches allow a high inertia load to be started with a smaller electric motor than would be required if where directly connected. Clutches are commonly used to maintain a constant torque on a shaft for tensioning of webs or filaments. A clutch may be used as an emergency disconnect disk to decouple the shaft from the motor in the event of a machine jam. In such cases a brake will also be fitted to bring the shaft (machine) to a rapid stop in an emergency. To avoid injuries to worker in a production system, when some enclosures (or door) are open or when the operator touches some critical element (panic bar); clutch–brake combinations can be provided. Under normal circumstances, the brake is disengaged and the clutch is engaged. During emergency situation, power fails and brakes are engaged. Clutches and brakes can be classified in a number of ways, by means of action (mechanical, pneumatic hydraulic, electric and automatic), by means of energy transfer between elements (positive mechanical contact and frictional), by the character of the engagement (square jaw, spiral jaw, and toothed, etc).

8.2 Clutches

Clutch is an important part of automobiles. Clutch is also finding its applications in many of the machines. Clutch is a mechanical device, which is used to connect or disconnect the source of power transmission system at the will of the operator. When a motor of a textile machine is to be started, the drive to the machine should be disconnected till the motor attains the full speed. Clutch may be regarded as a device employed to connect or disconnect shafts during their relative motion or when at rest. For the smooth performance of the driven shaft, both driving and driven shafts should be perfectly coaxial.

There are two types of clutches: (1) jaw clutch and (2) friction clutch.

8.3   Jaw/toothed clutches

A jaw clutch used in a carding machine is shown in Fig. 8.1. This consists of two disks, which have jaws/teeth that can fit with each other. The jaw disk 1 is fastened to the driving shaft A. The driving shaft gets its motion from the doffer through bevel gears. The other disk 2 is fitted on the driven shaft B and can be shifted along the driven shaft over a guide feather key by a shifting mechanism C. Both the shafts are coaxial. The driven shaft transfers motion to the feed roller through bevel gears. When jaws are engaged, motion is transmitted by direct interference between projections on two parts of the jaw clutch. Disengaging the clutch can disconnect the drive to the feed roller. These clutches can only be engaged at low speeds (relative velocity 60 rpm for jaw clutches, 300 rpm for toothed clutches). For high-speed application, the engagement will cause violent shock and noise.

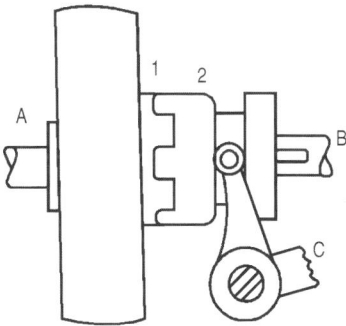

8.1 A jaw clutch.

Jaw clutches were also used in looms. In conventional blow room, the drive to pedal roller of piano-feed regulating mechanism is effected through jaw clutch. Once the lap attains the pre-set length, a lever from the lap stop mechanism disengages the clutch and hence the drive to the pedal roller of the last beater of the blow room is disconnected. Toothed clutches are used for under winding of yarn in ring frame. When a ring cop is full, the clutch disengages and the ring rail goes to bottom position.

The advantage of these clutches is positive engagement and, once coupled, can transmit large torque with no slip. They are sometimes combined with a friction type clutch, which drag the two elements to nearly the same velocity before the jaws or teeth engage. This is called synchromesh clutch. Jaw clutches are comparatively smaller in size for same power transmission.

The satisfactory operation of jaw clutch depends on the accurate machining of two halves to ensure perfect contact between all conjugate jaws. The correct mounting of two halves on the respective shafts is essential for correct engagement.

8.4   Friction clutches

Friction clutches are gradually engaging clutches. Driving shaft may be rotating at full speed while the driven shaft either stationary or rotating at much lower speed is brought into connection with the former. As the engagement of the clutch proceeds, the speed of the driven shaft attains the speed of the driving shaft. The torque transmitting efficiency of friction clutches depends on the frictional force between two bodies which are pressed together. To increase the friction between the two bodies, a special material, 'friction material', is provided on one of these bodies.

8.4.1   Friction clutch with two-plane discs

This consists of two discs or flanges or plates as shown in Fig. 8.2. One disc 'driving disc' is fastened to the driving shaft. The driven disc is free to move along the driven shaft due to splined connection. Both the shafts are coaxial. During disengagement of the clutch, a contact lever keeps the driven disc away from the driving disc. For the engagement of the clutch, the contact lever has to be gradually released. Then a spring provides an actuating force to the driven disc forcing it to move towards the driving disc and finally makes contact with it. The driven disc starts rotating at low speed due to the friction between the discs. When the contact lever is fully released by a hydraulic cylinder, the spring provides the required axial force to press the driven disc against the driver disc, the friction force between them increases, and the driven disc attains the speed of the driver disc. Torque is transmitted by means of frictional force between these plates. Disc clutches are used in the drive from calender roller to lap rollers of a lap forming machine. When a lap attains its preset length, clutch disengages. Due to braking action, calender stops feeding web. However, lap rollers run for about a metre severing the lap.

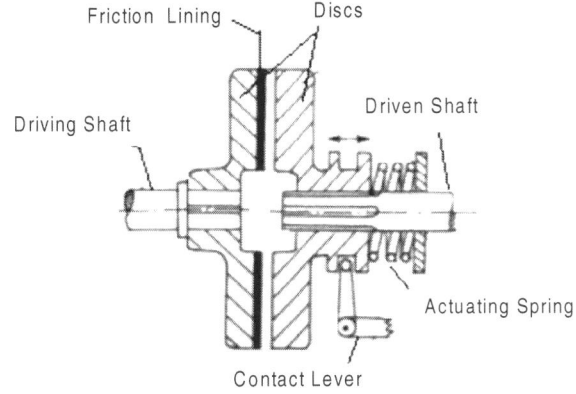

8.2 Single plate friction clutch.

The friction clutches are classified as two-plane discs or multiple-lane discs depending upon the number of friction surfaces. Based on the shape of the friction lining, they are classified as disc clutches, cone clutches or expanding shoe clutches. Friction clutches permit smooth engagement at any speed. In any event of over loads, the friction clutches slip momentarily, safeguarding the machine or mechanism against breakage. Two-plane disc clutches are used where there is enough radial space.

8.4.2   Friction clutches with multi-lane discs

A multi-plane disc friction clutch is shown in Fig. 8.3. It consists of two sets of discs A and B. A set of driven discs, 'A', are mounted to the driven shaft by means of splined sleeve, so that they are free to move in an axial direction. A specially shaped plate is fastened to the driving shaft. It has a rim and an L-shaped face or drum. When the driving shaft rotates, the drum also rotates along with the driving shaft. Radially eqi-spaced axial holes (three or four) are provided on the rim and the drum, so that a bolt can pass through each set of the holes.

The driving set of discs 'B' is also made with radially equi-spaced holes. The bolts pass through the holes of the drum, driving discs, 'B' and the rim of the plate. A clearance fit between the bolts and the holes in the driving disc allows the discs B to move in an axial direction. The bolts are rigidly fixed to a revolving drum. Normally, the discs 'A' are placed compressed under spring force, so that they pressed against the driving discs, 'B', and torque is transmitted to the driven shaft. For disengagement of the clutch, contact levers move the driven discs away from the driving ones. Hardened steel and hardened bronze are used to make the driven and driving discs respectively.

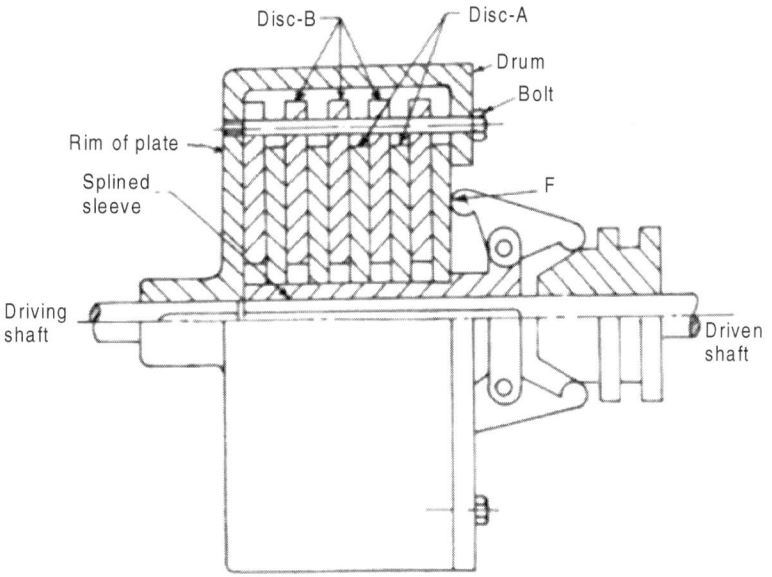

8.3 Multi-disc friction clutch.

8.5   Materials for friction lining

Asbestos-based materials and sintered metals are commonly used for friction lining. There are two types of asbestos friction discs: woven and moulded. A woven asbestos friction disc consists of asbestos fibre woven with endless circular weave around brass, copper or zinc wires and then impregnated with rubber or asphalt. The endless circular weave increases the centrifugal bursting strength. Moulded asbestos friction discs are prepared by moulding the wet mixture of brass chips and asbestos.

The woven materials are flexible, have higher coefficient of friction, conform more readily to clutch surface, costly and wear at faster rate compared to moulded materials.

Asbestos materials are less heat resistant even at low temperature. Sintered-metal friction materials have higher wear resistance, high temperature-resistant, constant coefficient of friction over a wide range of temperature and pressure, and are unaffected by environmental conditions. They also offer lighter, cheaper and compact construction of friction clutches.

8.6   Cone clutches

The cone clutches are simple in construction and are easy to disengage. However, the requirement for the coaxiality of the driving and driven shafts is critical. A cone clutch consists of two working surfaces, viz., inner and outer cones, as shown in Fig. 8.4. The outer cone is fastened to the driving shaft and the inner cone is free to slide axially on the driven shaft due to splines. A helical compression spring provides the necessary axial force to the inner cone to press against the outer cone, thus engaging the clutch. A contact lever is used to disengage the clutch. The inner cone surface is lined with friction material. Due to wedging action between the conical working surfaces, there is considerable normal pressure and friction force with a small engaging force. The semi-cone angle 'a' is kept greater than a certain value to avoid self-engagement; otherwise disengagement of clutch would be difficult. This is kept around 12.5°. Cone clutch is used to delay the start of drafting unit of a ring frame.

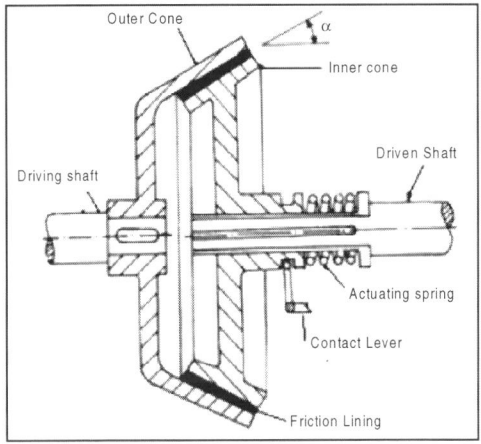

8.4 Cone clutch.

8.7   Centrifugal clutches

Whenever the load on the driver (motor) has to be engaged after the driver has attained its full speed or a critical speed, a centrifugal clutch may be used in those situations. The centrifugal clutch permits the drive motor to start, warm up and accelerate to the operating speed without load. Then the clutch is automatically engaged and the driven element is smoothly brought up to the operating speed. These clutches are very much useful for heavy loads where the motor cannot be started under that load. They are widely used in textile machinery. For example, centrifugal clutches are used in drive from motor to cylinder and lickerin of modern card. Drive from motor to motor-pulley is through this clutch. Once the motor attains the required speed, the centrifugal clutch engages, transmitting the drive to lickerin and cylinder, thus safe guarding the motor during start-up.

The centrifugal clutch works on the principle of centrifugal force, which increases proportionally to the square of the rotational speed. A centrifugal clutch is shown in Fig. 8.5.

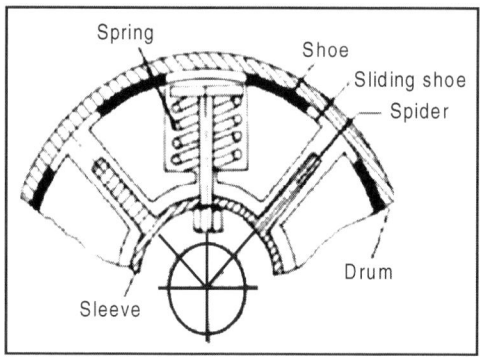

8.5 Centrifugal clutch.

Spiders are mounted radially, equi-spaced on the driver or input shaft. They form the radial guides or spaces around the driver shaft. In each guide, a sliding shoe is retained by a spring.

The outer surface of the sliding shoe is provided with a lining of friction material. The entire assembly of the spider, shoes and spring is enclosed in a co-axial drum, which is mounted on the output or driven shaft.

As the rotational speed of input shaft increases just after the start-up of the input drive, the centrifugal force acting on the sliding shoe increases. This causes each shoe to move in radically outward direction. The shoe continues to move with increasing speed until they contact the inner surface of the drum, overcoming the spring force. Torque is transmitted due to frictional force between the shoe lining and the inner surface of the drum. The centrifugal force is corresponding to speed. Just before the engagement of shoe with the drum, the centrifugal force is equal and opposite the spring force. Therefore, above the engagement speed, the centrifugal force on the shoe is slightly more than the spring-force, the shoe overcomes the spring force and contacts the drum.

8.8   Brakes

A brake is a machine element which is used either to stop the machine or retard the motion of a moving system, such as a rotating rollers or drums or vehicle where the driving force has ceased to act or is still acting. In practice most brakes act upon drums mounted on the driving shafts or driven shafts. In such cases brake will act either upon the internal surface or external surface of the drum. The brakes acting on the brake drums do not make contact along the whole periphery and the part making contact with the drum is called shoe.

The shoe has to expand for internal contact and close in for external contact. When the braking action takes place, the energy absorbed by the brake shoe is converted into heat energy and dissipated to surroundings.

Heat dissipation is a serious problem in brake applications.

Depending on the shape of friction material, the mechanical brakes are classified as disc type, drum brakes. In addition, hydraulic, magnetic, pneumatic and eddy current brakes are available.

8.8.1 Mechanical brakes

(a) Block brake with short shoe

A block brake consists of a rotating drum (brake drum) against which a brake shoe (or block) is pressed by means of a pivoted lever is shown in Fig. 8.6. The friction force between the shoe and the brake drum acts against the direction of rotation of the drum at the contact region. This causes retardation of the drum. When the friction force is very high, the drum stops rotating. The angle of contact, 0, between the shoe and the brake drum is usually kept less than 45°, for uniform intensity of pressure between them. The main disadvantage of the block brake is the tendency of the drum shaft to bend under the action of normal force (N).

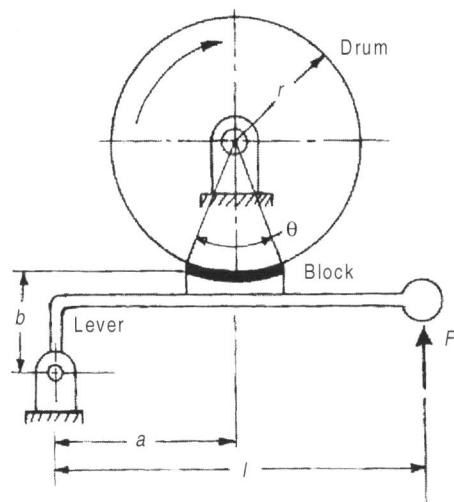

8.6 Simple block brake.

Single block brake with short shoe is used in scutcher to give pressure to the lap during winding. As the lap builds-up, the lap-rack tends to move upwards. A brake drum is mounted on to the shaft that carries the pinions meshing with the rack. As lap diameter increases, the pinion rotates due to the movement of the rack. Since a brake shoe mounted on a lever with adjustable weight is in contact with the drum during lap winding, a braking action is always effected against the upward movement of the lap-rack.

This helps in building a compact lap. These brakes were commonly used in winding of sliver-laps and ribbon-laps.

(b) Pivoted double block brake

Pivoted double block brake is used in braking the spindle for piecing the yarn in ring frame.

A spindle brake is shown in Fig. 8.7. In this case the tendency for unseating the brake is eliminated.

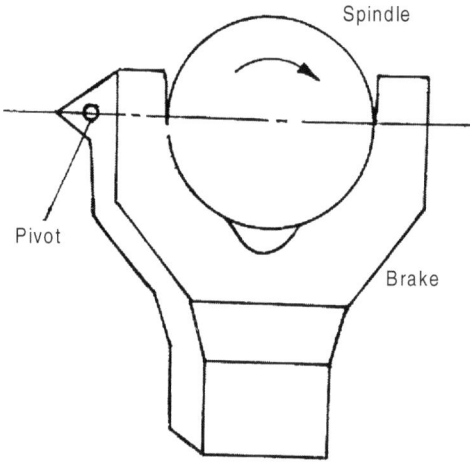

8.7 Ring spindle brake.

(c) Internal expanding brake

An internal expanding brake is shown in Fig. 8.8. It consists of a shoe, which is pivoted at 'A' and on the other end 'B' an actuating force F acts. A friction lining is provided on the shoe.

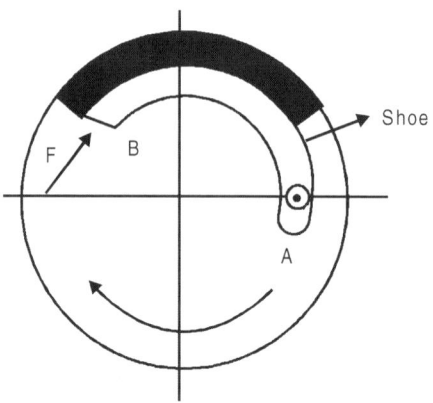

8.8 Internal expanding brake.

The complete assembly of shoe with lining and pivot is placed inside the brake drum. Under the action of the actuating force the shoe contacts the inner surface of the drum. Internal shoe brakes, with two symmetrical shoes, are used in lap forming machine and TFO. The actuating force is usually provided by a hydraulic cylinder. When a lap is about to reach its preset length, braking action on the main shaft slows down the machine.

(d) Band brakes

In a band brake, a flexible steel band lined with friction material, presses against the rotating brake drum. Figure 8.9 shows a simple band brake, where one end of the steel band passes through the fulcrum of the actuating lever (O). The other end of the band is connected to the lever at point (A) a distance 'a' from the pivot point. Actuating force is applied at point (B) on the lever. The working of steel band is similar to that of a stationary flat belt on rim of a pulley. Therefore, the ratio of tensions on the steel band is given by

$$T_t/T_s = e^{\mu\theta}$$

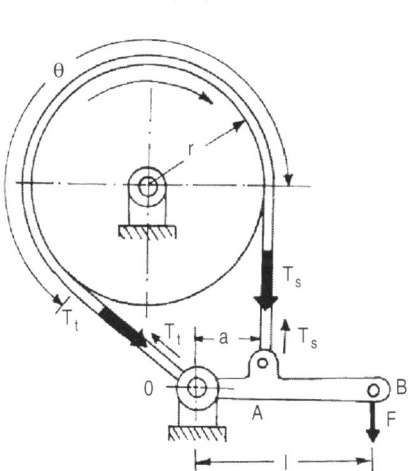

8.9 Simple band brake.

The torque Mt absorbed by the brake is given by

$$Mt = (T_t - T_s)\, r$$

Where,

Mt = torque capacity of the brake (N-mm);

r = radius of the brake drum (mm)

Considering the forces acting on the lever and taking moments about the pivot (0),

$$T_s.a = Fl$$
$$F = T_s.a \,/l$$

(e) Differential band brake

A differential band brake is one where both the ends of the band are not passing through the pivot of the actuating lever. Differential band brakes are designed for the condition of self-locking. These breaks were used earlier in sliver and ribbon lap machines to apply pressure on lap spindle during lap formation. Currently the pressure on the lap spindle is applied by pneumatic means.

Considering forces acting on the lever and taking moments about the pivot, we can get,

$$(Fl) + (T_t\, b) - (T_s\, a) = 0$$
$$F = (T_s\, a - T_t\, b) \,/\, l$$

Substituting equation in the above expression,

$$F = T_s\, (a - b\, e^{\mu\theta}) \,/\, l$$

For self locking condition,

$F = 0$ or negative, $a < = b\, e^{\mu\theta}$

Therefore, the condition of self-locking is given by $\% < = e^{\mu\theta}$

Similarly, when the drum rotates in counter clockwise direction, the actuating written as,

$$F = Ts\, (a\, e^{\mu\theta} - b)/l$$

9.1   Introduction

The sley performs the beat up motion and lays the last pick of weft inserted to the fell of cloth. The detail mechanics of the sley is dealt in this chapter. The eccentricity ratio is a critical factor to be considered for different weaving machines. The ratios depend upon the width and type of machine. In some machines the sley is operated by crank mechanism and in others by cam, as in the case of unconventional machines. The merits and demerits of higher sley eccentricity ratios are highlighted,

9.2   Sley displacement or motion

9.2.1   Beating-up

Pushing the currently inserted length of weft, known as the pick, into the previously formed fabric at a point known as fell is the third and last in the sequence of primary motions of a weaving machine called loom.

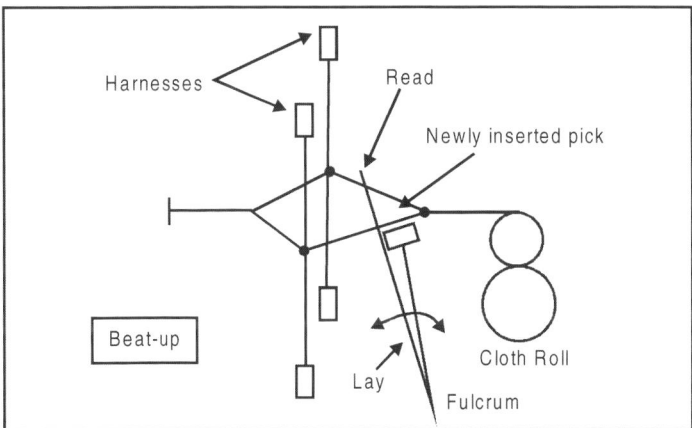

9.1 Sley and beat-up motion.

The sley reciprocates for the reed to push the weft into the fell of cloth.

Two sley swords extend down from the race board to a fulcrum point called as the rocking shaft. The sword pin connects both the sley swords with a crank arm just below the race board. The other end of the crank arm fastens round the bend in the crank shaft called as crank.

As the crank shaft revolves, the crank arm and hence the top end of the sley are made to reciprocate with a displacement which approximates to simple harmonic motion. It should dwell as long as possible at the rear dead center to leave the largest possible section of the angle for the weft insertion. On the other hand, it should beat-up the weft to the fabric fell strongly in order to obtain the desired pick density.

9.2.2   Loom timing

The timings of most of the events in the loom cycle are governed by the position of the reed and thus the sley.

For example, the reed must be on its way towards the back of the loom before the shed is large enough to admit the shuttle. This determines the timing of the picliing mechanism which is directly related to the position of the reed and sley.

Some others are related to it indirectly. For example, the timing of the weft break stop motion is related to the flight of the weft carrier, which is governed by the position of the reed.

The timings on the weaving machine are stated in relation to the angular position of the crankshaft (main shaft) which operates the sley.

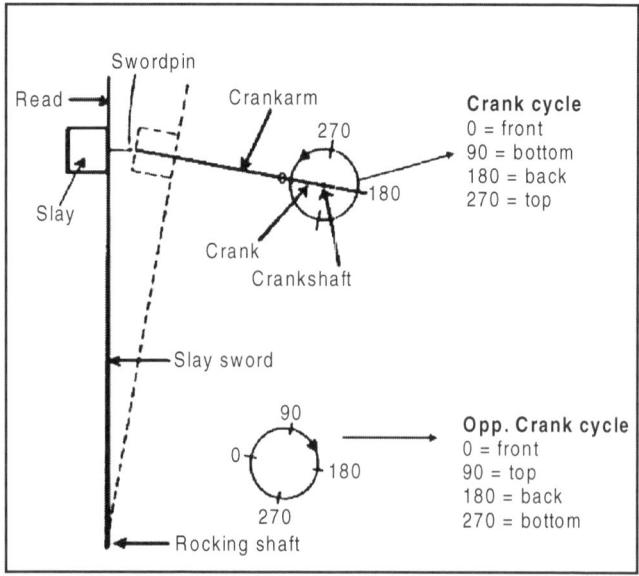

9.2 Loom timing.

The path traced out by the axis of the crank pin is called the 'crank circle'.

The arrow on the crank circle shows the usual direction of rotation of the crankshaft. When the crank and crank arm are in line, the sley is in its most forward position. The crank circle is graduated in degrees from this point in the direction of rotation of the crankshaft. Any timing can be stated in degrees, as, for example, 'heald level at 300°'.

Looms are provided with a graduated disc on the crankshaft and a fixed pointer to make settings in relation to the angular position of the crankshaft. With the reed in its most forward position, the disc is adjusted so that the pointer is opposite to 0° on the graduated scale. The loom may then be turned to any desired position manually, the disc turning with it and the pointer remaining vertical and indicating the angular position of the crank shaft. In modern looms with microprocessors, the main shaft position is displayed on a screen, but the setting principle remains same.

9.2.3   Types of beat-up mechanisms

There are several types of mechanisms used for achieving the required motion of sley. They are mainly divided into two:

1. Link-type beat-up mechanisms
    • Four-link
    • Six-link
    • Multi-link
2. Cam operated beat-up mechanism, special mechanism.

Factors affecting the motion of the sley

• When the sley is operated by crank and crank arm, its motion approximates to simple harmonic.
• The extent to which it deviates from simple harmonic motion has practical significance and is governed by the following factors:
    (a)   the radius of the arc along which the axis of the sword pin reciprocates,
    (b)   the relative heights of the sword pin and crankshaft, and
    (c)   the length of the crank in relation to that of the crank arm
• *The normal arrangement:* The axis of the crankshaft is on a line passing through the extreme positions of the axis of the sword pin, and the reed is vertical at beat-up. (Fig. 9.3)

The sword pin travels along an arc of a circle centered upon the rocking shaft. This modifies the movement of the sword pin and hence of the reed, but, since the radius of the arc (length of the sley sword) is large (about 0.75 m); thus, the effect is small enough to be neglected.

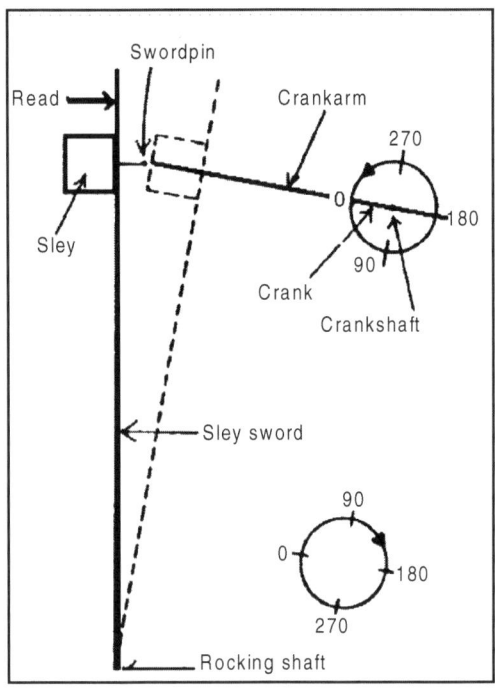

9.3 Reed, sword pin, crank arm and crank position during beat-up.

- Relative heights of the sword pin and crankshaft
- Raising or lowering the crankshaft from its normal position affects both the extent and the character of the motion of the sword pin.
- 'Moving the crankshaft 10 cm up or down from its normal position;
- increases the distance moved by the sword pin by about 8%.
- increases the sword pin's velocity as it approaches its most forward position and to decrease it as it approaches its most backward position.
- the result is increasing the effectiveness of beat-up and of allowing more time for the passage of the weft carrier.

Sley eccentricity, e = r/l

- The ratio **r/l**, where *r* is the radius of the crank circle and *l* is the length of the crank arm, is called the sley eccentricity ratio, *e*.
- The larger it is, the greater is the deviation from simple harmonic motion. (Fig. 9.4)
- If '**e**' increases then,
    - more time is available for shuttle passage
    - more effective beat up (beat up force increases) but mechanical problems occur on loom parts.

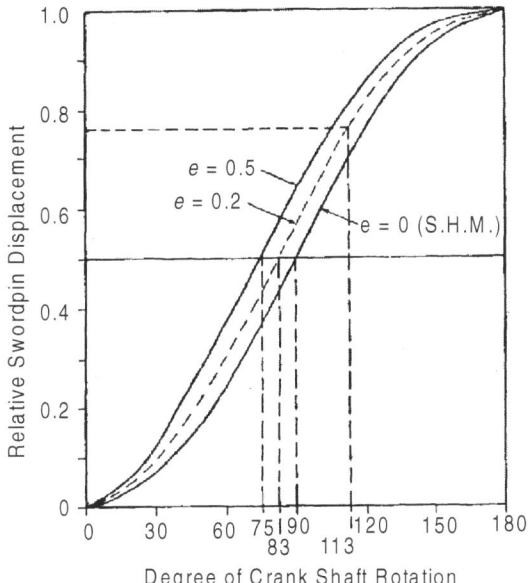

9.4 Motion of the sley.

- The displacement of the sword pin is expressed as a fraction of its total displacements for a half revolution of the crankshaft.
- Crankshaft is in the normal position.
- The curves for the second half will be the mirror images of those for the first half.
- With simple harmonic motion (corresponding to $e = 0$ and indefinitely long crank arms), the sword pin attains its maximum velocity and exactly half its maximum displacement at 900 and again at 2700.
- With a finite value of e, with any arrangement possible in Practice if the sley is crank-driven, the sword pin attains its maximum velocity and half its maximum displacement earlier on its backward movement and later on its forward movement.

Eccentricity ratio (e)	Positions of crankshaft at half maximum displacement	Period during which displacement is at least half maximum
0.0	90° and 270°	180°
0.2	83° and 277°	194°
0.5	75°and 285°	210°

- *As the sley-eccentricity ratio increases*
    - the sley remains longer nearer its most backward position, and more time is available for tile passage of tile shuttle.

- increases tile maximum attainable velocity of the sley around beat-up.
- *We may summarize the advantages of a high sley-eccentricity ratio as follows:*
  (a)   it facilitates the passage of the shuttle; and
  (b)   it tends to increase the effectiveness of beat up.
- *The effects of altering the sley-eccentricity ratio within the practicable limits are, however, greater than those obtained by altering the height of the crank shaft.*

The disadvantages of a high sley-eccentricity ratio:

- A high value implies rapid acceleration and deceleration of the sley around beat-up.
- It increases the forces acting on the sword pins, crank pins, cranks, crank arms, crank shaft, and their bearings, indirectly on the loom frame.
- A high sley-eccentricity ratio will therefore demand more robust loom parts and a more rigid loom frame in order to prevent excessive vibration and wear, so that, for a given standard of performances the loom will cost more.

For the above said reason, most loom makers tend to avoid eccentricity ratios greater than about 0.3. However, there are exceptions.

Loom maker	Type	r (cm)	l (cm)	e = r/l
Saurer	Cotton, tappet	6.25	15.0	0.42
Rüti	Cotton, dobby	7.60	33.5	0.23
Picanol	Cotton, tappet	7.20	32.4	0.225
Prince (water-jet)	Rayon, tappet	3.33	22.9	0.145
Dobcross	Worsted dobby	8.90	43.2	0.21
Northrop	Industrial blanket, tappet	10.80	20.3	0.54

In general, the forces involved in accelerating and decelerating the sley will be proportional to the effective mass of the sley and the square of its velocity.

- For given sley eccentricity ratio, its velocity will be proportional to the product of the loom speed and the length of the cranks.
- A is a typical automatic loom for weaving cotton and spun rayon fabrics and B is a non-automatic loom for weaving heavy woolen industrial blankets.

Type	Reed space (m)	Speed (picks/mm)	Length of crank (cm)	Mass of sley
Loom A	1.14	220	7.0	M
Loom B	5.33	65	10.8	M

Cam operated mechanisms (for most of the shuttleless weaving machines)

- To achieve high loom speeds on shuttleless weaving machines;
  - the mass of the sley should be reduced to a minimum
  - the distance through which it reciprocates should be as low as possible.
- In order to minimize the weight of the sley, heavy parts associated with picking are mounted on the loom frame except some means of guiding for the weft carrier through the shed.
- Since the device used to carry the weft through the shed will have a smaller cross-section than a shuttle, a smaller shed and hence a smaller sweep of the sley will be sufficient. (Fig. 9.5)

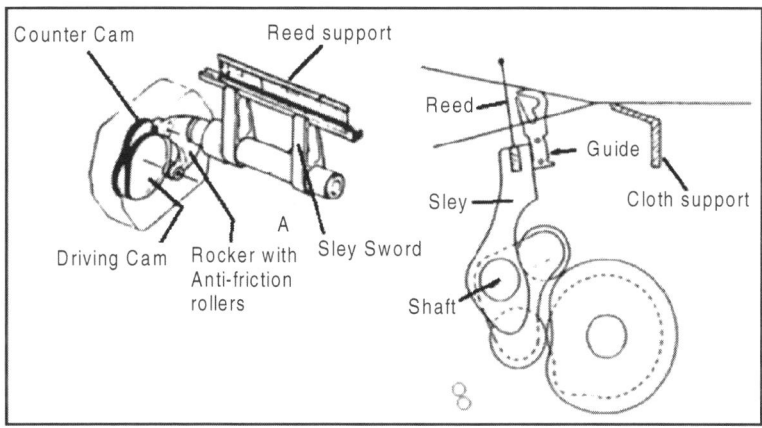

9.5 Cam drive for sley – Sulzer projectile weaving machine.

- The picking mechanism is mounted stationary on the machine frame, then the sley must dwell in its most backward position during the whole of the time occupied by weft insertion.
- Only a cam mechanism can precisely ensure the dwell position within the required range of 220° to 250°.
- Sulzer sley drive uses several pairs of matched cams, spaced at intervals across the width of the sley.
- The motion of the sley, including its period of dwell, is positively controlled.
- In most models of the Sulzer weaving machine, the whole of the sley movement is completed in 105° of the weaving cycle, which thus allows the sley to dwell in its back position for 255°.
- In the narrow, single-color machines, which run at the highest speed, the sley movement is spread over 1400 to prevent excessive vibration, which leaves a dwell period of 2200. This is acceptable because the

weft carrier does not travel over a long distance as in the wider machines.

- The cam mechanism gives the outstanding advantage of a possible cam change for various working widths. But the manufacture of this mechanism is very exacting; only a minimum clearance is admissible between both cams with rollers to avoid impacts in the mechanism.
- Moreover, this mechanism occupies a large space in the warp wise direction and makes, therefore, the arrangement of the other weaving mechanisms rather difficult. (Fig. 9.6)

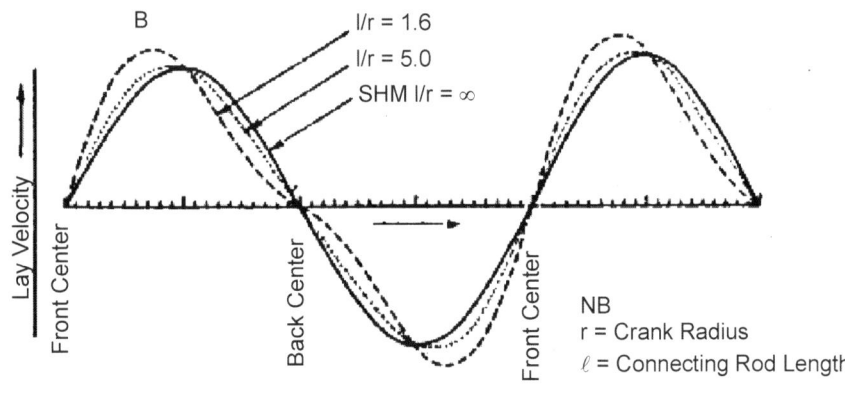

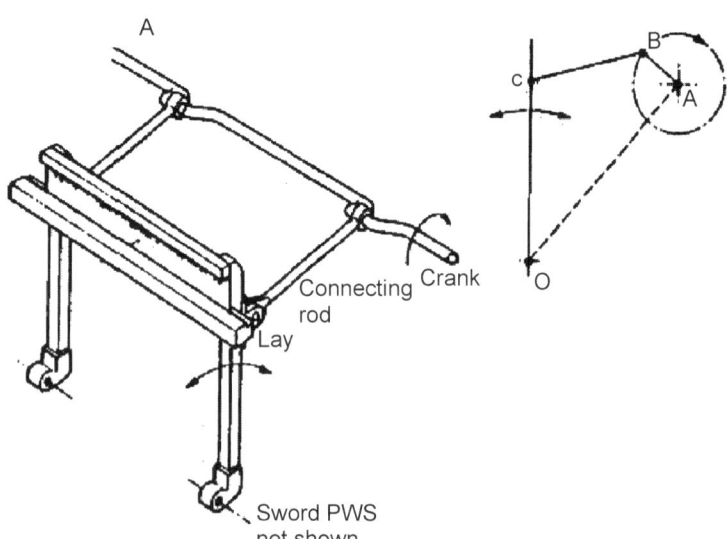

9.6 Sley mechanism and its velocity.

Effect of various connecting rod lengths

l /r	Connecting Rod	Type of Movement	Type of Fabric
Greater than 6	Long	Very smooth with low acceleration forces	Fine cotton, silk, continuous filament
Between 6 and 3	Medium	Smooth	Medium density cottons
Less than 3	Short	Jerky with high acceleration forces	Heavy cottons, woolen

Forces in beat-up process

- Beat-up force
- Warp and fabric tensions
- Weaving resistance (Fig. 9.7)

Beat-up force (F):

- The force exerted by the reed onto the warp and cloth system during beat up.
- The beat-up force must overcome the resistance of the warp ends under tension, open in front of the penetrating weft thread.
  - the frictional resistance between ends and pick as the pick is pushed through the warp sheet. (Fig. 9.8)

Transmitted impulse of force:

- Different types of beat-up mechanisms have different abilities to transmit adequate impulses of force to the weft beat-up.
- The impulse of force generated by the reed = mass × speed
- Sley must also be able to transmit this quantity of motion to the fabric fell.

It is consumed for the beat-up.

- The supplied impulse is dependent on the beat-up angle in which the reed is in contact with the cloth fell.
- The beat-up angle is adjusted automatically by increasing the beat-up zone so that equilibrium is attained between the supplied and consumed impulse of force.
- Increase in no. of picks → cloth fell tends to advance against the direction of the reed beat-up → increase beat-up zone and beat-up angle (beat-up force).

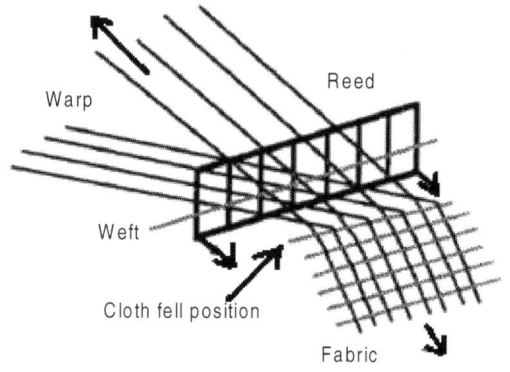

9.7 Forces in beat-up.

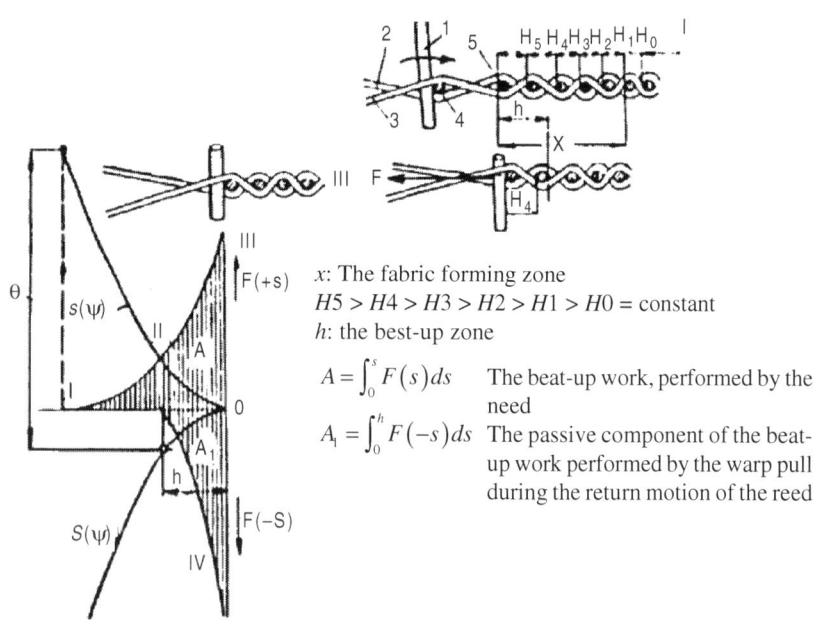

x: The fabric forming zone

$H5 > H4 > H3 > H2 > H1 > H0 = $ constant

h: the best-up zone

$A = \int_0^s F(s)\,ds$ The beat-up work, performed by the need

$A_1 = \int_0^h F(-s)\,ds$ The passive component of the beat-up work performed by the warp pull during the return motion of the reed

9.8 Beat-up force.

$$\text{Beat up force/end} = 2\mu\alpha T_2$$

μ – the coefficient of friction between ends and picks.

α – determined by the crimp levels, the yarn dimensions and pick spacing, the dimensions *m* and *l*.

T_2 – the tension in the warp is affected by the beat-up, increases with the displacement of cloth fell and, the beat-up force is also a function of T_2.

Example 1: Analytical method

In a crank and connecting rod mechanism (Fig. 9.9), the lengths of the connecting rod and crank are 1200 mm and 300 mm respectively. The crank is rotating at 180 rpm. Find the velocity of the reed, when the crank is at an angle of 45°, with the horizontal.

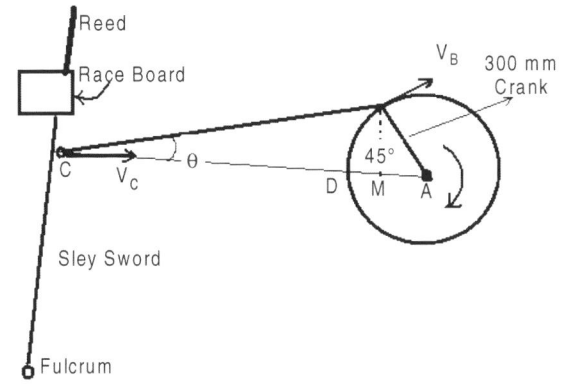

9.9 Crank and connecting rod mechanism.

Solution:
Given: r = 300 mm = 0.3m; l = 1200 mm = 1.2 m; N = 180 rpm; θ = 45°
∴ Angular velocity of the crank,

$$\omega = \frac{2\pi \times 180}{60} = 6\pi \text{ rad/s}$$

From the geometry of the figure, we find that

$$Sin\phi = \frac{BM}{BC} = \frac{ABSin45°}{BC} = \frac{0.3 \times 0.07}{1.2}$$

$$= 0.1768 \text{ or } = \tau = 10° \text{ 11'}$$

We know that velocity of the reed,

$$V_c = \omega (l \text{ Sin } \tau + r \text{ Cos } \theta \text{ Tan } \tau)$$
$$= 6\pi (1.2 \text{ Sin } 10° \text{ 11'} + 0.3 \text{ Cos } 45° \text{ Tan } 10° \text{ 11'})$$
$$= 6\pi [(1.2 \times 0.1768) + (0.3 \times 0.70 \times 0.196)] \text{ m/s}$$
$$= 4.72 \text{ m/s.}$$

Example 2. Velocity diagram method

The lengths of connecting rod and crank are 1125 mm and 250 mm, respectively. The crank is rotating at 420 rpm. Find the velocity with which the sley will move, when the crank has turned through an angle of 40° from the inner dead centre.

Solution: l = 1125 mm = 1.125 m; r = 250 mm = 0.25 m; N = 420 rpm; θ = 40°

We know that velocity of crank,

$$\varpi = \frac{2\pi N}{60} = \frac{2\pi \times 420}{60} = 44 \text{ rad/s and velocity of B,}$$

$$V_B = \omega \, r = 44 \times 0.25 = 11 \text{ m/s.}$$

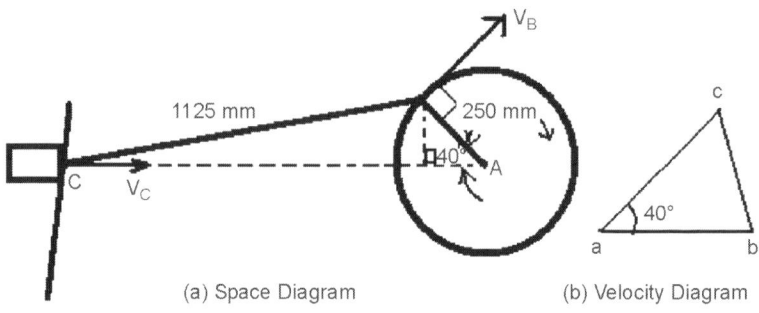

| (a) Space Diagram | (b) Velocity Diagram |

9.10 Space and velocity diagram.

First draw the space diagram as shown in Fig. 9.10 and then draw the velocity diagram also as discussed below:

1. Take some suitable point 'a' and draw a horizontal line representing the direction of motion of sley, i.e. Vc.
2. Through 'a', draw another line 'ab' representing the direction of motion of B i.e. V_B which is at 40° with the horizontal.
3. Now cut off 'ab' equal to 11 m/s to some suitable scale.
4. Through b, draw 'bc' perpendicular to the connecting rod.
5. Now ac gives the velocity of piston to the scale. By measurement, we find that velocity of sley.

$$Vc = ac = 8.3 \text{ m/s}$$

10
Mechanics of shedding motion in weaving

10.1 Introduction

This chapter deals with the mechanics of the shedding motion in a loom. It involves calculation of the depth of warp shed, the various types of heald movements, the geometry of shed and shed angle. These calculations are practically significant, since the depth of shed requires the shuttle height to be known which in turn is decided by the size of the weft package used. It is useful to find out the depth of shed for a given size of shuttle.

10.2 Calculation of depth of shed

It is necessary to know the depth or height of the shuttle to be used when measured along its front wall; the distance of its front edge, at the time it enters the shed, from the cloth fell and the distance from the cloth fell to the heald controlled by that particular tappet. It is also necessary to know the distance from the treadle lever fulcrum to the bowl centre, and the distance between the bowl centre to the point of heald connections. All these are shown in Fig. 10.1.

The shed depth also depends on the sweep of the sley. Referring to the figure, the depth of shed or as is known the lift of heald could be determined by use of similar properties of triangles. By ascertaining the depth of the shed, it will also lead to estimation of the lift of the tappet.

Referring to Fig. 10.1, let

A = cloth fell to the front of the shuttle wall
B = cloth fell to the front heald
C = height of the shuttle front well
x = depth of front heald movement
z = displacement of treadle lever connected to the front heald
a = treadle lever pivot to bowl centre
b = treadle lever pivot to front heald connection
y = lift of tappet of front heald

Thus, we can write,

$$X = B.C/A, \text{ and from the construction}$$

we find, x = z, so that lift of tappet is

$$a/y = b/z = b/x,$$

Putting the value of x from the first relation, we have

$$Y = (a.b.c)/(A.b)$$

Similarly, the lift of tappet of the second heald can also be determined.

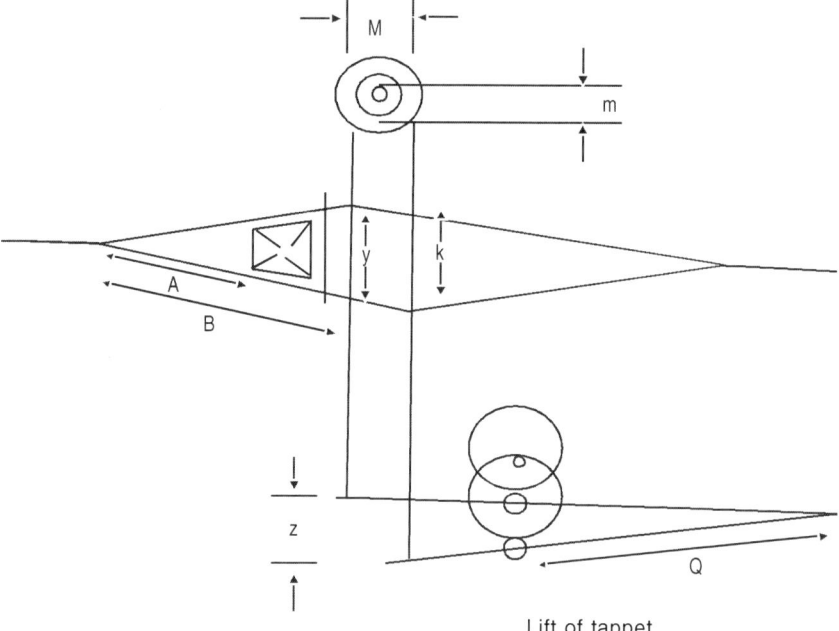

Lift of tappet

10.1 Calculation of depth of warp shed in loom.

The tappets being negative in nature, their action in actual practice is similar to the positive types. This can only happen if the top of the healds are connected with top revering system (here rollers) by means of inelastic straps. To satisfy the condition, the relative diameters of the reversing rollers must be found to keep the tappets always in contact with treadle bowls so that no loss in lift takes place.

Thus if,

m = roller diameter for front held, and
n = roller diameter for second heald,

We have,

$$m/n = x/k, \; n = (m.k)/x = (a.m.k)/(y.b)$$

Thus, roller diameter for the back heald will be greater as the value of b will be lesser, so the rollers are in direct proportion to the depth of the shed concerned.

10.3   Heald movement

The warp lines during shedding can be shown by means of displacement diagrams as shown in Figs. 10.2(A), (B) and (C). The type of movement is approximately simple harmonic and this is plotted vertically against time intervals in degrees of crank shaft rotation.

In Fig. 10.2(A), the curve shows the displacement of the warp line for one pair of tappet with dwell of 120°. The shed closes at 270°.

The three curves in Fig. 10.2(B) are for three pairs of ordinary tappets, and having a dwell of 120°. They are set at intervals of 20° which would be abnormal in actual practice but is made use of to indicate the effect on the actual dwell of the complete shed. Curve A is for a pair of tappets set to cross at 270°, curve B is for tappets set 20° in advance of A, whilst curve C is for tappets set 20° later than A. Thus the three pairs of heald staves would reach the centre position at intervals of 20°. The effect of this is that the effective dwell is reduced to 80°.

Figure 10.2(C) is the displacement diagram of three pairs of especially designed tappets. Each pair reaches the closed shed position at 270° but at different heights in the shed. Curve A shows the heald staves crossing at the centre of the shed, curve C for the second pair crossing at about 1.5 cm below the centre line, while curve B is for a pair crossing at about 1.5 cm above the centre line. This setting is considerably more than that used in practice. The complete dwell is, however, still 120°.

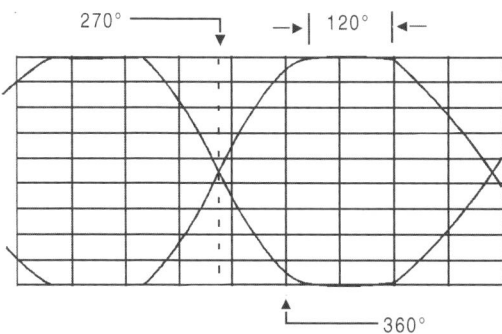

10.2(A) All threads crossing at same height and time.

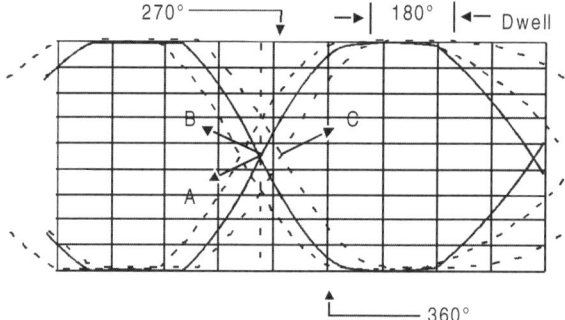

10.2(B) Healds crossing at same height but different time.

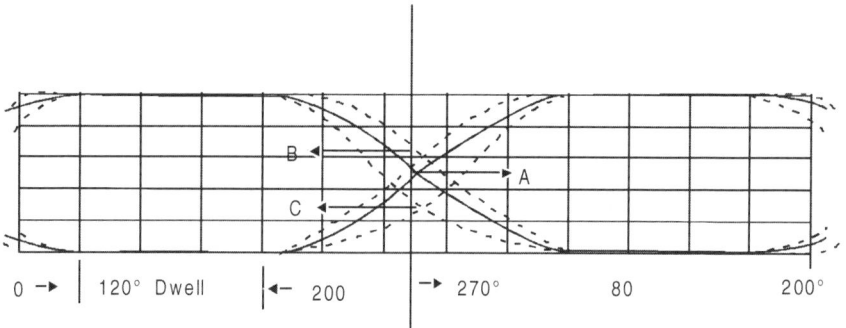

10.2(C) Healds crossing at same time but different in height.

10.4   Geometry of warp shed

10.4.1 The size of the shed

The width and depth of the shuttle are dependent on the diameter of the weft package that will go into the shuttle. In considering the size of the shed required for a given size of shuttle, the important dimension is the depth of the shed at the front wall of the shuttle.

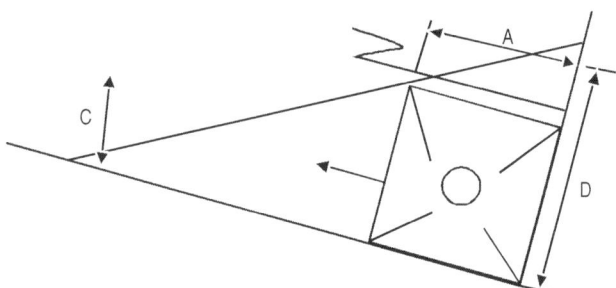

10.3 Geometry of warp shed.

In Fig. 10.3, C is the shuttle height and A is the width, B is the distance from the cloth fell to the reed, D is the depth of the shed at reed. Much bending or interference can be tolerated. During the flight of the shuttle both B and D varies because of the motion of the reed and heald unless the shuttle passage coincides with the period of dwell.

Figure 10.4 shows the plotting of the depth of shed at the shuttle front against the angular position of the crank shaft for a particular weaving machine. Curve B, which is asymmetrical, was obtained with the healds staves set to cross at 0°. If it is assumed that the shuttle enters the shed at 110° and leaves at 240°, the depth of the shed at the front of the shuttle as it enters and leaves will be given in Table 10.1.

Table 10.1 Depth of shed at shuttle front during entry and exit

	Depth of shed (cm)	
	Entering	Leaving
Curve A: Healds crossing at 270°	2.44	0.94
Curve B: Healds crossing at 0°	2.36	2.54

The depth of the shed at the front of the shuttle, expressed as a fraction of the height of the shuttle front wall, is called "bending factor" or "interference factor". This can also be expressed as percentage interference as

$$\text{Interference percent} = \frac{\text{depth of shed at shuttle front}}{\text{height of the shuttle front wall}} \times 100$$

It indicates the extent to which the warp threads are deflected, if at all, by the shuttle. An interference or bending factor less than 1.0 (i.e. 100%) implies deflection. And when the factor is greater than 1.0 it would mean undue strain in the warp threads. If we take the height of the front wall of the shuttle, in Fig.10.3 in a machine, to which the figure relates, to be 2.8 cm the interference factor will be as arc given in Table 10.2.

Table 10.2 Interference factor of shuttle

	Interference factor	
	Entering	Leaving
Curve A	0.87	0.34
Curve B	0.84	0.90

Two extreme cases are shown in Fig. 10.4, in which the dotted lines represent the position the top shed would occupy if it were not deflected by the shuttle. In this particular case, there would be some deflection of the warp line by the shuttle on entering and leaving with the both shed timing. The amount of bending, however, is quite small except when the

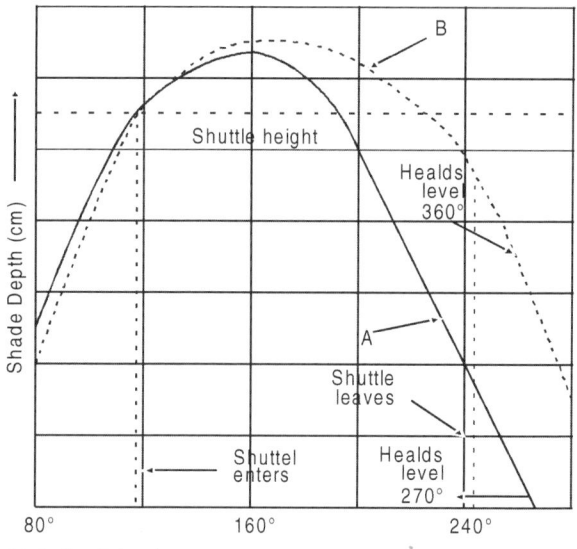

10.4 Shed depth curve.

shuttle is leaving with healds set to cross at 270°. The question arises as how much bending or interference can be tolerated.

Referring to the Fig. 10.5, we see that for both curves bending occurs as the shuttle enters only between 110° and 120°.

Warp Defelection

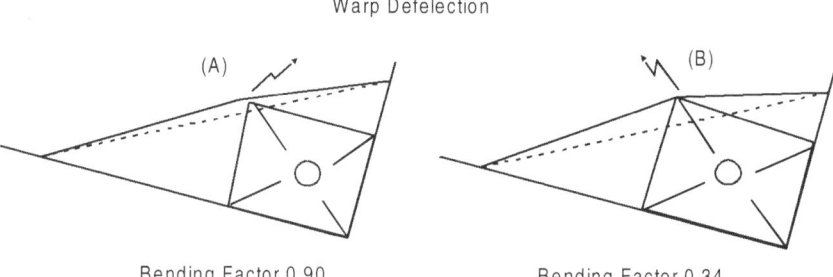

Bending Factor 0.90   Bending Factor 0.34

10.5 Deflection of warp.

For curve B, healds crossing at 0°, the depth of the shed is greater than the height of the shuttle between 120° and 230°. A small amount of bending does occur between 230° and 240° as the shuttle leaves the shed. These conditions would be acceptable even in weaving low twist continuous filament yarns. If required, a slight increase in the depth of the shed would be desirable to eliminate bending.

Curve A represents conditions that could not be tolerated in weaving continuous filament yarn but quite common in weaving spun cotton yarns. There is no more bending as the shuttle enters than with curve B, but the shuttle would deflect the warp about 195° until it leaves the shed at 240°. By the time it leaves, the bending factor is 0.34, as shown in Fig. 10.4, and it is clear that there must be quite severe rubbing of warps by the shuttle during the later part of its passage through the shed. This can normally be accepted in weaving spun yarns in order to get the benefit already mentioned, better warp cover and more effective pick packing.

10.4.2 Shed angle

The shed depth or shed angle depends on the movement of the sley. The movement of the reed is slightly greater than that of the sword pin because the distance R from the rocking shaft to the reed is greater than that of to the sword pin, K, as shown in Fig. 10.6. The shed opening diagram at the shuttle front can be used to determine the earliest or latest crank shaft position for the shuttle being in the shed. From these, the time for the shuttle passage, from the time its front end enters the shed at one side until its rear clear the shed at the other end, can be estimated. The amount of shed opening at the shuttle front wall can also be estimated.

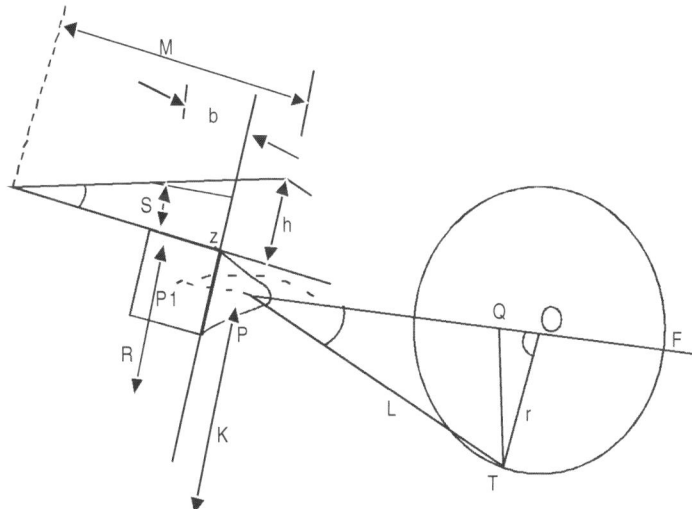

10.6 The shed angle.

From the above figure, we find that the displacement of the sword pin from the 'Beat-up' position is

$$= PP_1 = OP_1 - OP = OP_1 - (QP + OQ)$$

Hence,

$$PP_1 = (n + 1) - (1 \, Cos \, \theta + r \, Cos \, \theta) \qquad \ldots (1)$$

It will be better to express Cos β in terms of θ. In triangles PQT and QOT, the side QT is common.

Therefore,

$$QT = 1 \, Sin \, \beta = r \, Sin \, \theta$$

So that, r Sin

$$sin \, \beta = (r \, sin \, \theta)/1$$

Squaring, we find:

$$sin^2 \, \beta = (r^2 \, sin^2 \, \theta)/1^2 \qquad \dots (2)$$

By trigonometry,

$$Cos^2 \, \beta = 1 - sin^2 \, \beta,$$

Putting the value of $sin^2\beta$ from (2) in the above relation, we obtain,

$$Cos^2 \, \beta = 1 - (r^2 \, Sin^2 \, \theta)/1^2,$$

and extracting the roots

$$Cos \, \beta = \sqrt{[1 - (r^2 \, Sin^2 \, \theta)/1^2]} \qquad \dots (3)$$

The right hand side of Eq. (3) can be expanded binomially, and neglecting powers higher than two, because their effect

$$Cos \, \beta = 1.()/21^2] \qquad \dots (4)$$

Now we have transformed Cos β in terms of θ and, by putting the values in Eq. 1, we have

$$PP_1 = (r + 1) - [1\{1 - (r^2 sin^2 \, \theta)/21^2\} + r \cos \theta]$$
$$= r + 1 - 1 + [(r^2 sin^2 \, \theta)/21] - r \cos \theta$$
$$= r \{1 - \cos \theta + [(r^2 sin^2 \, \theta)/21]\} \qquad \dots (5)$$

We find the displacement of the reed from the cloth fell to be

$$= XZ = PP_1(R/K),$$

And the displacement of the shuttle front wall from the cloth fell is

$$= XY = [PP_1(R/K)] - b,$$

The shed opening at the shuttle front is

$$S = XY \tan \infty$$

So,

$$S = \{[PP_1(R/K)] - b\} \tan \infty \qquad \dots (6)$$

Putting the value of PP_1 from Eq. (5) in Eq. (6) we get,

$$S = [\{r (1 - \cos \theta + (r/21) sin^2 \, \theta) R/K\} - b] \tan \infty \qquad \dots (7)$$

Again,

$$h/2 = M \tan\infty /2,$$
$$\infty/2 = \tan^{-1} (h/2M),$$

Therefore,

$$\infty = 2 \tan^{-1} (h/2M)\text{degrees} \qquad \dots (8)$$

In Fig. 10.7, we shall find the relative positions of cloth fell, reed, healds, warp and the shuttle almost the same way as shown in figure. From the data given it would be easy to find out the shed angle, shed height and stroke of the tappet. The crank shaft has brought the sley into its back centre position, and in doing so the sley has moved through 120°. The angle between the race board and the reed is 87°, and the angle CFD between shed line and the horizontal is therefore 15°, as is shown in Fig. 10.7 (b), and remembering that sum of the angle of triangle is 180°. The front of the shuttle is 11.5 cm from the fell F, and its height is 3.75 cm. A clearance of 0.5 cm is required between the top front edge of the shuttle and the top shed line.

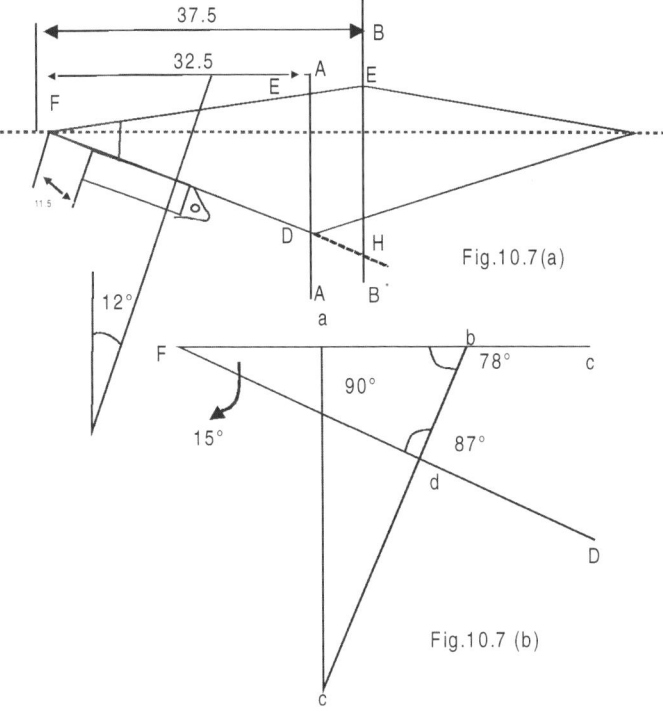

Fig.10.7(a)

Fig.10.7 (b)

10.7 Shed geometry

The full shed angle, EFD, is given by

$$\tan EFD = (3.75 + 0.5)/11.5$$

$$= 4.25/11.5 \text{ from which angle EFD is } 20° \text{ 17'}$$

Subtracting 15° from 20° 17', we find the angle EFC to be 50° 17', the angle between the top shed line and the horizontal. We can now consider how much vertical movement is required by the healds staves AA and BB. The movement required by the stave AA is sum of the distances CD and CE. We thus have in triangle CDP;

$$CD/FC = \tan 15°,$$

from which,

$$CD = 32.5 \times 0.2679 = 8.71 \text{ cm};$$

again in triangle EFC;

$$CE/PC \tan 5° \text{ 17'},$$

from which,

$$CE = 3cm$$

the total movement of the stave is, therefore $8.71 + 3 = 11.71$ cm

A similar calculation could be made to determine the stave BB, which is equal to the sum of GK and HK, but since the triangles EDF and GHF are similar

$$GH/ED = 37.5/32.5$$

And

$$GH = 11.71 \times (37.5/32.5) = 13.2 \text{ cm}$$

In order to obtain the necessary shed angle, the staves AA and BB have therefore to have movement of 11.71 and 13.2 cm, respectively. The assumption has been made that the warp threads move exactly in sympathy with the healds, but there is a little difference due to the size of the heald eye. If an allowance of 1 cm is made for heald eye depth, the final stave movements become: stave AA = 12.75 cm; and stave BB = 14.2 cm.

The generator of the heald stave movement may be a tappet, and between the tappet and the stave there is a linkage with a particular ratio. Let us assume that this ratio is 1.8. Then, for a tappet stroke of 1 cm, the stave will move 1.8 cm.

The stroke of the tappet for the front stave AA will be

$$12.71/1.8 = 7.1 \text{ cm (approximately)}$$

For the stave BB, the ratio could be slightly less, depending on the machine design, say 1.5. The tappet stroke for the stave BB is therefore,

$$14.2/1.5 = 9.4 \text{ cm (approximately)}$$

with other types of shedding device, a similar approach would be used.

Loom tappet drives

11.1   Introduction

Tappets are driven from the crankshaft directly or indirectly by means of spur gear wheels. In most of the weaving machines the centre distance between the gear wheels being fixed the ratio of the crank and tappet shaft drive is maintained in such a manner that the ratio between the two are equal to the number of picks per round. There may be difference in the number of sum of the teeth in the wheels, but such difference shall not exceed more than one tooth, and in all cases the wheels should be so chosen that the ratio is maintained to give correct number of picks to the round. In case of Ruti-C machines, where the positive tappets are situated outside the machine framing, different gear wheels both for crank shaft and tappet shaft are used to get the proper drive ratio, as shown in Table 11.1.

Table 11.1 Wheel combinations for different picks

Picks	Crank shaft wheel	Tappet shaft wheel
4	24 T	96T
5	20 T	100T
6	17 T	102T
7	15 T	105T
8	20 T	160T
9	18 T	162T
10	16 T	160T

It will be observed that sum of the teeth in two wheels has remained same up to 7 picks, and then again the total number has increased for 8 and 9 picks but in both cases they are same. For 10 picks the total is entirely different and this would be possible only when the pitch circle diameter has been altered. Thus if one wants to make a quick change in the cloth structure a large storage of different wheels would be a must.

11.2   Different types of drives

In majority of the ordinary non-sophisticated weaving machines the number of teeth in the crankshaft wheel is half the number in the bottom shaft or low shaft wheel on which generally the tappets of 2-picks weave repeat are mounted, and this is in conformity with the position of the picking tappets. But when the picks per repeat are 3 or 4 or even more, then the large wheel on the low shaft cannot be changed to accommodate the proper wheel. The reason being that the centre distance between the two shafts is generally fixed. Thus, for anything greater than 2 picks per repeat, the tappets must be mounted on what is known as countershaft or tappet shaft.

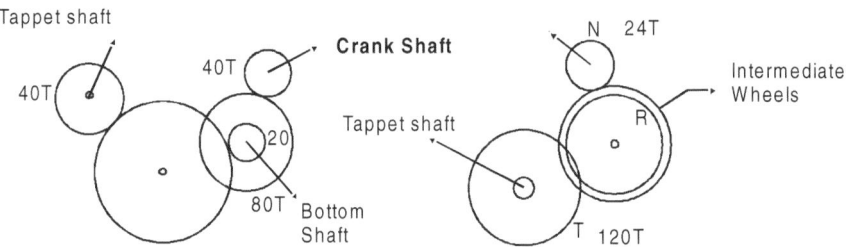

11.1 (a) Tappet drive with idler wheel; (b) Tappet drive with no idler wheel.

Figure 11.1 shows the tappets which are placed under the healds and are mounted on the tappet shaft, flow, for every revolution of the crankshaft the tappet shaft will create, for-4 picks repeat design, which is given by

$$(1 \times 40 \times 20)/(80 \times 40) = 1/4 \text{ revs.}$$

Thus the relation between the crankshaft and the tappet shaft rotation will be in the ratio of 4:1. In other words, for every rotation of the tappet shaft, the crankshaft will rotate four times. So, it is seen that the rotation of the crankshaft will be equal to the number of picks per design repeat. The carrier wheel is omitted in the calculation as it works as an idler wheel to convey the rotation in the opposite direction, and if the space available is not suitable for the two main wheels to mesh.

It is customary for the tappet wheel to contain a number of teeth which is multiple of the weaves produced. Here the tappet wheel has 120 teeth which is divisible by any number up to and including 10. But this number is not divisible by 7 and 9. The number of teeth required in the change wheel N, as shown in Fig. 11.1, for weaves which are complete on 2, 3, 4, 5 and 6 picks per repeat, is obtained by dividing the particular number into the number of teeth contained in the tappet wheel and in that case the intermediate wheels Q, and R will be replaced by a single carrier wheel. For example, if we have a design repeating on 6 picks it will require a change wheel of

$$(120/6) = 20 \text{ teeth}$$

But for tappets intended for weaving 7, 8, 9 and 10 picks, the two wheels Q and R will be required in place of a single carrier wheel. This is due to the reason that neither 7 nor 9 will divide 120 into a whole number. And, if 8 and 10 picks are intended they will render the wheels calculated to be useless as they will be too small, i.e. wheel numbers would be respectively 15 and 12 which will not just fit in. Now, by adopting a small wheel on the crankshaft with 24 T, the intermediate wheels required would be in inverse proportion as 24 multiplied by the number of picks per repeat of the tappets to be used is to the number of teeth in the tappet wheel 120. For a 8-pick pattern the wheel ratio is found to be (24 × 8)/120 = 192/120. But these wheels being too large, wheels of same ratio will have to be used. Thus the wheel Q with 48T and wheel F with 30 T give the same ratio and can be conveniently used.

For various pick-pattern the following wheel ratios could be used with the relation.

$$(P \times N \times R)/(Q \times R) = \text{rpm of the tappet shaft,}$$

where,

p = number of picks per repeat,
N = crankshaft wheel,
Q = Intermediate driven wheel,
R = intermediate driving wheel, and
T = tappet shaft wheel.

For 8-pick pattern, the ratio Q/R is found to be 192/120 = 96/60 = 48/30.
For 7-pick pattern the same ratio is found to be 168/120 = 84/80 = 42/30
For 9-pick pattern the ratio is 216/120 = 108/60 = 54/30.
Thus, from the above figures we can conclude that with given number of wheels in the figure, the generalized relation becomes

$$Q/R = (6\ p)/30$$

It must also be remembered that though the ratio may be kept same, but the number of teeth in the wheels should not be too small and prevent any meshing of the wheels. The wheel ratio must be kept within the limitations of the machine designer.

11.3   Drive without idler wheels

Hanton takes an approach which is normally found in the ordinary weaving machines is illustrated in Fig. 11.2. Figure shows the usual arrangement of gearing between the crankshaft and the shafts on which shedding tappets are mounted. Since one pick is inserted in the cloth for every rotation of

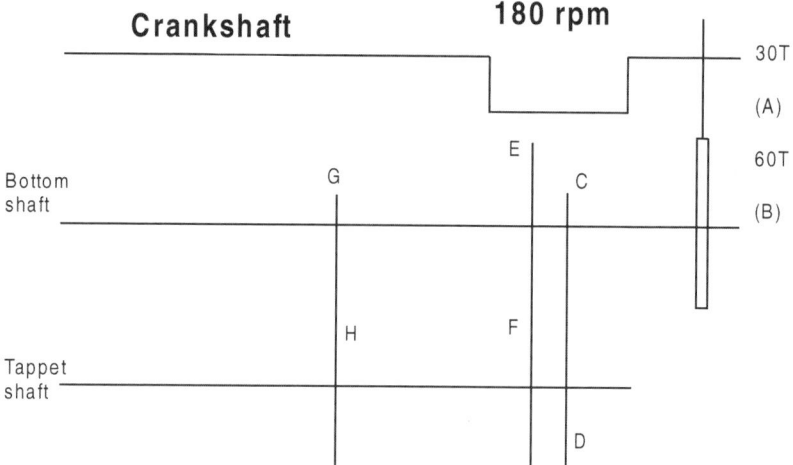

11.2 Tappet drive.

the crankshaft, the speed of the crankshaft divided by the speed of the tappets shaft must be equal to the number of picks per repeat of design.

For 2-picks design, the tappets are placed on the bottom shaft.

If the speed of the crankshaft is 180 rpm, the bottom shaft will have a speed of

$$(180 \times 30)/60 = 90 \text{ rpm},$$

provided the crank and bottom shaft wheels have 30 T and 60 T, respectively. Or, in other words, the bottom shaft will rotate at half the speed of the crankshaft. Similarly, the bottom shaft will rotate one-third of the crankshaft speed when weaving a 3-pick design. Whereas, in case of 2-pick repeat it is possible to put shedding tappets on the bottom shaft, it is impossible to do so for a 3-pick design without changing the bottom shaft wheel, and this too will not be possible unless the pitch of both crank and bottom shaft wheels are changed, or provisions for changing the centre distance between these two shafts are incorporated in the machine. Generally none of the above approaches are made but for certain weaving machines as was described in Table 5.1. This problem leads us to the concept of mounting tappets on another shaft, known as counter or tappet shaft, to be driven from the bottom shaft.

From Table 11.1, it was seen that under certain conditions the sum of the teeth in both the wheels are kept constant with a tolerance of only one tooth. Similarly, for drive to the counter shaft the sum of the countershaft driven wheel and the bottom shaft wheel (being driver) teeth must be constant with a tolerance of one tooth only. This facilitates algebraic relations to be used in calculating the teeth in the wheel for the required purpose. Hence, for a 2-pick design, referring to Fig. 11.2, we can write

$$(A/B) = 1/2, \text{ or } B = 2A, \text{ and}$$
$$A + B = 90,$$

so that we have 2 A + A = 90,
from which we get A = 90/3 = 30.

Thus, wheel A, the crankshaft wheel, will have 30 teeth, and the bottomshaft wheel B to have 60 teeth, keeping the ratio undisturbed and the sum of the wheel teeth unchanged.

For a 3-pick design, as stated earlier, the tappet will be mounted on the countershaft. The countershaft with D wheel is driven by wheel C on the bottomshaft and whose sum is 70. So we obtain

$$C + D = 70,$$

and from the drive relation we find

$$(A \times C) / (B \times D) = 1/3,$$

So that, $(30 \times C)/(60 \times D) = 1/3$, and c = (2D)/3.
Putting the value of C in the earlier relation, we have 5D = 210, or D = 42, and C = 28, and the speed ratio of 2/3 is maintained.

Again for 4-pick design, we have
F + F = 70, and the drive ratio as earlier would be

$$(30 \times E)/(60 \times F) = 1/4, E/F = 1/2.$$

and E = F/2
so that as before F = 140/3 = 46.67, and E = 23.33.

As fractional wheels are inconceivable, the number must either be increased or decreased to give a whole number, and at the same time the speed ratio must not be sacrificed. But, as the difference of one tooth does not make difference in transmission of motion, we can use a 23 T wheel of E and a 46 T of F, and at the same time the ratio of 1/2 is maintained.

For 5-pick design the wheels are found to be 50 T for wheel H, and 20 T for G. The above trains of wheels are capable of weaving up to 5-pick designs; for 6- and 7-pick designs other combination of bottom and crankshaft wheels must be found. For 6-pick design, if the total number of teeth is taken to be C + D = 80, then D = 60 T and C = 20 T. Again for the 7-pick weave with C + D = 90, we shall have D = 70 T and C = 20 T. Under this condition the pitch of the wheels will have to be so adjusted that the centre distance between the two shafts are not altered.

11.4   Movement of hooks, knives and warp threads in jacquard shedding

It is evident that when a hook is engaged by on the knife it cannot be disengaged while the load is resting on the knife, nor would it be possible to reengage the hook unless the knife moved below the hook position at the same stage in its motion, so that the knife entered the hook from below. To achieve this a bottom board is placed in such a position that the hook rests on it at a point in its downward path which is above the lowest point to which the knife descends. Figure 11.3 shows the position of the knives plotted against time. The position below which the hooked member cannot travel due to the presence of the bottom board is also plotted on the graph. If a hook is engaged by a knife it will move down with it but will leave the knife when point (A) is reached, and it cannot be reengaged until the knife reaches point (B). During the time the knife moves from (A) to (B) a new selection can be made, if desired; if the new card has a hole opposite the needle operating this hook, the hook will again rise with the knife after it passes point (B). If a blank is present on the card the hook will not be engaged and will remain in down position resting on the bottom board.

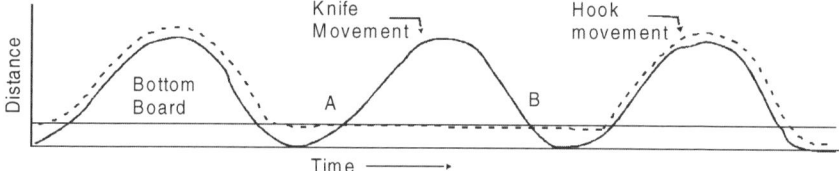

11.3 Movement of hooks and knives in a single acting single cylinder jacquard.

While some of the objections to the simplest jacquard system have been overcome, it can be seen that the requirement of a stationary oven shed has still not been met. A further development which uses two hooks and knives to operate each warp thread makes this possible. In double-lift double cylinder machine the two sets o knives are made to move up and down by two four-bar linkages, so that each performs a simple harmonic motion which repeats every two cycle of the weaving machine. The two motions are, however, out of step with each other, which plots the movement of each knife against time. Also shown on this diagram is the limiting lower position of the two hooks, which is determined by the fixed position of the bottom board.

Figure 11.4 shows the movement of the warp thread when threading is employed, the dashed line showing the warp thread displacement for the following section cycle both hooks are selected at the same time by the movement of the needles but only one hook selection is effective bees-use only one of the hooks can be engaged by the knife (3). The other hook, which can only engage with knife (4), cannot he engaged because knife (4) is descending at this stage. By this selection, therefore, the warp is lifted until the hook reaches position (A) in the figure, knife (3) begins to descend. During this cycle the position of the shed for the next cycle is

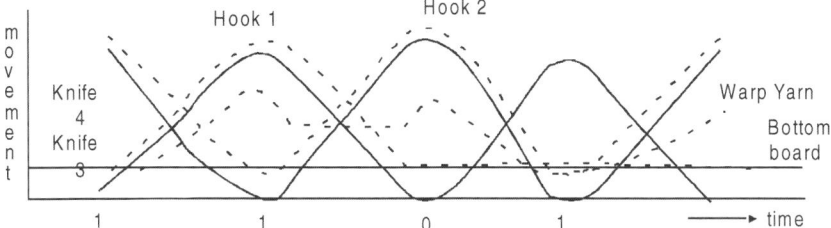

11.4 Movements of hooks, knives and warp in double acting double cylinder jacquard.

chosen. In the sequence shown in Fig. 11.4 the shed is to be opened again. When selection takes place only one hook is resting on the board, namely, that which can engaged with knife (4). Having been selected it is engaged by knife (4) on its upward travel, and consequently the shed is fully opened again at (c). On the next cycle, the hook resting or, the board is not selected and the other hook descends with knife (4) until it too, rests on the board. The shed therefore remains closed. The movement of the hooks is shown by the dotted lines in figure. The warp thread, whose displacement is the average of the two hook displacement, is also shown in the Fig. 11.4 and it can be seen that the effect of this selection mechanism and threading arrangement is that the warp thread is maintained in its raised position throughout the first two cycles. This produces the required open-shed action.

The action is clearly desirable on the mechanical ground. It can be seen that the resulting action is smooth, and that no sudden impulsive forces act on the knives, such as occur in all other methods. It must, however, be noted that the thread movement is only half as great as the knife movement, and this may result in an undesirably large knife movement in some instances.

12
Analysis of shuttle movement in weaving machine

12.1   Introduction

This chapter analyses the movement of the shuttle in relation to that of the reed. The shuttle movement is traced during the various degrees rotation of the crank shaft. The time of entry and exit of shuttle through the warp shed are also discussed. The crucial factors governing the shuttle movement are highlighted. The important settings for effective shuttle movement are also indicated. A mathematical relationship exists between the speed of the weaving machine, the length of shuttle traverse time required for the shuttle to traverse the distance, and the average velocity of the shuttle, and this has been elaborated at one part of the chapter.

12.2   Shuttle traverse

During one revolution of the crankshaft the sley recedes from the contact with the cloth fell at the front centre position at 0°. The sley then moves with more increasing velocity until the shed opening at the front of the reed is large enough to admit the shuttle at about 95° past beat-up; thereafter the sley slows down to zero at the back centre (180°). It again begins to move forward until it reaches the front centre (0°), to beat-up. As part of this cycle the shuttle is moving the sley, in contact with the reed and the lower sheet of warp and at the same time must move with the sley, maintaining contact between the reed and warp without being under the direct control of the machine mechanism,

The sley, supported by the sley swords, is secured to the rooking rail, moves in an arc on this centre, being driven from the crankshaft through crank and crank arm. The movement imparted to the sley, which will be discussed in the later section, is usually described as eccentric, distinguishes it from simple harmonic motion, and oscillates in a horizontal plane. The type of movement desired is one which accelerates rapidly from zero at the front centre, to allow the reed to move quickly away from the cloth fell and provide an opening for the shuttle to enter the shed at about 95°; at this point it is slowing down and, since the shuttle is moving in contact with the reed, etc. the retardation is also imparted to the shuttle, causing pressure between it and the reed.

At the back centre, the change of sley movement from back to front motion maintains this pressure and thus helps to steady the shuttle in its passage through the shed. The variable forward movement of the sley from 180° to 0° is not very important since the shuttle is in the box for the greater part of the time. The approximate displacement of the shuttle in its passage through the shed and due to the sley movement is shown in Fig. 12.1.

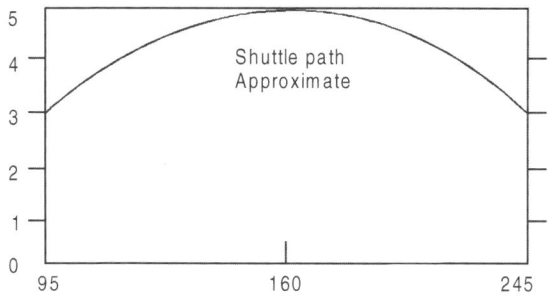

12.1 Path of shuttle traverse.

A smaller deflection, of similar type is caused by the vertical displacement of the sley,

being in it the highest position at the front centre and, falling during its backward journey, and is maximum at 180°.

In a 100 cm weaving machine, the speed of which is 200 picks per minute (ppm), the shuttle

traverse would enter the shed at 100° and leaves the shed at about 220° taking 120°, to move a distance of 125 cm, i.e., width of reed + effective length of shuttle.

12.3   Control of shuttle during weaving

The movement of the shuttle through the warp shed is partly controlled by the setting of the shuttle boxes and fittings, and partly by the sley construct on which the race board, usually made of hard wood, is secured. In many of the ordinary weaving machines the race board is made with slight down ward curve from each end towards the centre with about 1:180 gradient in a 110 cm reed space machine. A similar backward incline is given where it is in contact with the reed. This construction assists the shuttle in maintaining contact with the reed and the race board during its trajectory. The bottom warp line setting has a marked effect on the shuttle traverse in that if it is too high in relation to the race board the shuttle tends to rise against the top warp line near the end of its flight and be deflected back again, often causing warp damage. If the sitting is too low the yarn is damaged by friction against the moving race board.

The outer end of the box plate, i.e. the base plate on which the shuttle rests, should be set slightly higher than the inner end, to conform with the shape of the race board. The spindle is set higher and further from the box back at the inner end so that the picker rises as it moves along the spindle, thus lifting the rear end of the shuttle as it leaves the box, but if this movement is upward and excessive, the shuttle front will strike hard on the warp and may cause end breakage. However, this "on-and-off" setting, as it is called, should not be greater than 3 mm. The downward pressure on the rear end of the shuttle tends to lift it at the front and give a faulty shuttle traverse, and the usual method of correcting this fault is by raising the spindle stud.

The box front or fender is set wider at the inner end to conform to the setting of the spindle; the mount of clearance depends on the condition of the picker. With the close setting of the box front, the shuttle may become partially locked between the box back and front which causes a loss in machine speed and when the shuttle leaves the box it moves at a very great speed. Wide setting may also be detrimental to weaving in a sense that shuttle may rebound in the box resulting in loss of speed in the succeeding pick as the gap between the picker and the shuttle end is reduced.

The box front or fender is set wider at the inner end to conform with the setting of the spindle; the mount of clearance depends on the condition of the picker. With the close setting of the box front, the shuttle may become partially locked between the box back and front which causes a loss in machine speed and when the shuttle leaves the box it moves at a very great speed. Wider setting may also be detrimental to weaving in a sense that shuttle may rebound in the box resulting in loss of speed in the succeeding pick as the gap between the picker and the shuttle end is reduced.

A common setting of reed to the race board is 88° and the shuttle used should be made to fit this angle. However it is difficult to maintain this condition in practice, and in that case the angle of the shuttle may be greater than the reed angel which will cause a wedging action as shown in Fig. 12.2.

This tends to prevent vibration being set up as the shuttle is propelled. The movement of the sley helps to maintain contact between the shuttle and reed. When picking begins the sley, i.e. moving away from the fell of the cloth at maximum.

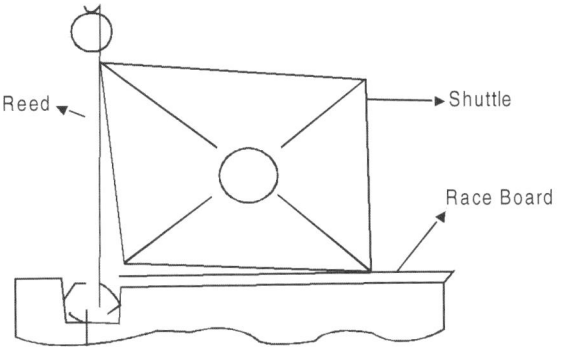

12.2 Setting of shuttle, reed and race board.

12.4   Shuttle flight and its timing

There is a simple numerical relation between the machine speed, the length of shuttle traverse time required for the shuttle to traverse the distance, and the average velocity of the shuttle. The traverse of the shuttle is equal to the width of the warp in reed plus the effective length of the shuttle. By effective length we would mean the flat part of the shuttle. This is bound to vary with the type and width of the weaving machine.

Then, if

R   =   the width of warn in reed (m.)
L   =   the effective length of the shuttle (m),
V   =   the average velocity of the shuttle (m/s),
θ =   degree of crankshaft rotation for shuttle passage,
P   =   speed of the machine (ppm)
t   =   the time available for the shuttle flight (s).

We have the-time for shuttle passage

$$t = \frac{60 \times \theta}{P \times 360}$$

and,

$$s = (L + R)$$

the average shuttle velocity is,

$$V = S/t = ((L + R)6P)/\theta$$

from which we obtain.

$$p = V \, \theta/6 \, (L + R)$$

Taking an example where the effective length of shuttle is 30 cm, the effective reed space is 120 cm with permissible average shuttle velocity of 12.5 m/s, shuttle trajectory taking 120° of crankshaft rotation. The permissible speed machine speed would be:

$$P = \frac{12.5 \times 120}{6(1.20 = 0.30)} = 167 \text{ rpm}$$

This speed is far exceeded in modern machines where it is worked at about 250 ppm. In increasing the machine speed greater than this imposes other limitations for which the speed cannot be exceeded. For a given machine width and effective length of shuttle the relation between P, V and Q takes the form $P \propto V \theta$. Thus if we want to increase the machine speed we shall have to increase either of the two variables or both. As the quantity V has a maximum limit, to which the shuttle speed could be increased, the only way out is to manipulate the time required for the passage of the shuttle through the warp shed.

It will be seen later that the entry of the shuttle in the shed is dependent on the movement of the sley. If the sley has an eccentricity ratio, which is defined at the ratio between the radius of the crank to the length of the crank arm, that is the ratio r/l, of about 0.2, the reed will have attained three-quarters of its total displacement. At that time the crankshaft position is about 113° on the machine timing cycle, and the shed is fully open and large enough for the shuttle to enter without excessive deflection of warp line. Thus it is seen that it is the displacement of the reed rather than the time of shedding which determines the earliest time of the shuttle entry (see Fig. 10.4). However, in practice, the shuttle enters the shed between 105° and 110°.

The latest time for the shuttle to come to rest in the shuttle box on the opposite side of the machine is determined by the monitoring device of the fast reed warp protection. When the machine is running normally the shuttle must be detected by the swell inside the box in time to cause any bang-off. If, for some reason, the shuttle fails to reach the box in time, the shuttle monitoring device of the fast reed protection must be able to stop the machine by 27 at the latest to avoid a shuttle trap. The monitoring device, the swell, must clear the stop rod dagger over the frog at about 260°, to allow for late arrival of shuttle due to any variation in shuttle flight time. As the swell is not fully displaced unless a few centimeters of the shuttle enters the box, the swell has to be contacted earlier. Say, for example, that a fast running loom with 200 ppm, the shuttle moves 0.075 m between contacting the swell and displacing it fully and that its average speed over this distance is 12.5 m/s, this will then require

(200 × 360 × 0.075) / (60 × 1.5) = 7.2° of crankshaft revolutions

On this basis, the latest time for the shuttle to contact the swell will be about 250°. Then the reed space is fully utilized the leading end of the shuttle would strike the swell as the trailing end just about clears the shed, so the shuttle rear must leave the shed at about 250°. Now the time available for the shuttle flight through the shed will be about 250° – 110° = 140°. If we take a look at Fig. 12.3, we should make the shuttle leave the shed a little earlier, if possible. There are certain yarns which are susceptible to abrasion, e.g. low twist continuous filament yarns; during weaving, it is desirable to delay the shuttle entry slightly. If again, the sley eccentricity is greater than what has been assumed in our example, it would be possible to have greater length of time for the shuttle flight.

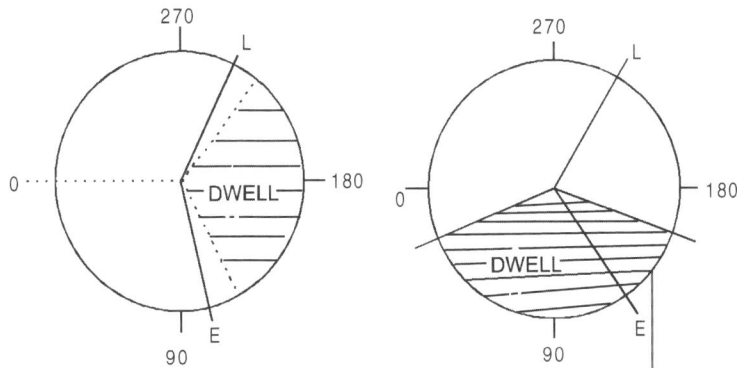

12.3 Timing of primary motions.

Thus we see that there is a distinct relation between the time of shuttle flight and the reed displacement, and certain limits of possible variation.

12.5   Weft insertion rate

The productivity of a weaving machine is dependent on the number of picks inserted in a given time and hence the rate of wet t insertion is dependent on the machine speed equation

$$P = (V\theta) / 6(L + R)$$

and that for given values of V and θ, the rate of weft insertion will increase with the width of the machine.

Figure 12.4 shows the weft insertion rate with the machine width, which have been calculated from the equation, assuming that the effective length of the shuttle is 0.30 m and its passage time is for 135° of the crankshaft revolution.

Now, if we weave a cloth on 2 m width machine instead of 1 m, the increase in the weft insertion rate would be 13.4%, this is also true for the area of the fabric produced. But by doubling the width from 1.5 m to 3 m, the increase falls off to the tune of 8.9% for wider machine. This is quite clear from Fig. 12.4.

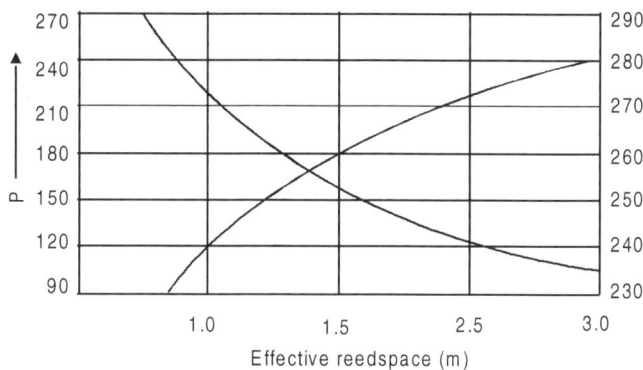

12.4 Rate of weft insertion.

For the garment makers it is always advisable to procure cloths of greater widths, as because wastages will be less, moreover, economy in dyeing and finishing will be of positive gain.

There is no doubt that wider machine will produce more cloth to terms of area. The wider machine will consume more yarn per unit time and the frequency of warp and weft breakages will rise meaning reduced efficiency. So, the calculated increase of 13.4% is not fully realized. Moreover, wider machines are costly, takes greater space but the number of machines will be less. With lesser number of looms the number of operatives will also be reduced. The economics of wider machines will have to be considered from all angles, e.g. initial cost, space labour, cost of re-equipment in preparatory and finishing, etc. The provision of weaving multiple width fabrics should be incorporated to utilize the width of the machine. These factors are coming in the way of introducing wide machines in industry.

13
Kinetics of shuttle picking

13.1 Introduction

The movement of the shuttle in a conventional weaving machine is rather complex and it took some years before it was understood fully from the works done by Vincont and Thomas, Vincent and Catlow. In order to get a proper understanding of the whole thing it is quite important to have a critical look from the very elementary aspects of the total system. Generally speaking, at every machine cycle, the shuttle is accelerated from rest to the maximum velocity within a distance of 20–25 cm at the commencement of picking cycle and then it travels with a slight retardation during its free flight through the shed and then again is decelerated within a short distance in the other shuttle box. Thus the whole thing can he divided into three parts which are not necessarily interconnected but each has some effect on the other.

These three stages can be described as

(i)   shuttle acceleration,
(ii) retardation during its free flight, and
(iii) shuttle checking

13.2 Shuttle acceleration

Vincent and Thomas during their experimental studies with an ordinary weaving machine were able to lot with some accuracy the shuttle displacement against angular position of the crank-shaft during picking and checking. The displacement curve for the shuttle during weaving is quite different from the curve plotted from observations made while the machine is turned over by hand. The later type of displacement was defined as the "nominal" displacement, and the former was called the actual displacement. The nature of this difference is shown in Fig. 13.1. Curves (A) and (B) relate to a non-automatic over pick machine. Curve (A) obtained is that of the nominal displacement, and the Curve (B) is the actual observed during actual running of the machine. The $0°$ on the graph is actually the 75° position of the crankshaft rotation when the picker and the shuttle are just about to begin to move. Comparing the two curves, it can be noted the nominal and actual displacement of the shuttle are the same at the start and again after about 30° of crankshaft rotation when the shuttle leaves the picker. Between 0° and 30° on the graph, there is a lag in the position of the shuttle and picker compared to that of the nominal position. The lag increases from 0° to 15°, where it reaches a maximum value as is indicated by the width of the loop. Thereafter, the lag decreases as the actual displacement gradually catches up with the nominal displacement, and finally overtakes it at about 30°, where the two curves intersect.

Then the machine is turned over slowly by hand, the force exerted on the shuttle by the picker is enough to overcome frictional resistance, which is due to pressure exerted by the swell on the shuttle. If, for example, the swell exerts a force of 65 N on the shuttle back wall, there will be similar and opposite force of 65 N between the box front and shuttle front wall. If the coefficient of friction between the shuttle and the swell, and between the shuttle and box front are both 0.25, the force exerted by the picker on the shuttle in overcoming this friction will be

$$2 \times 65 \times 0.25 = 32.5 \text{ N},$$

A similar force will be required to overcome friction when the machine is running, at speed; but, in addition, a much larger force will be necessary to accelerate the shuttle against the inertial resistance. Suppose, that the mass of the shuttle is 0.5 kg, and it is uniformly accelerated from rest to a speed 12.5 m/s over a distance of 0.2 m, then since, we have

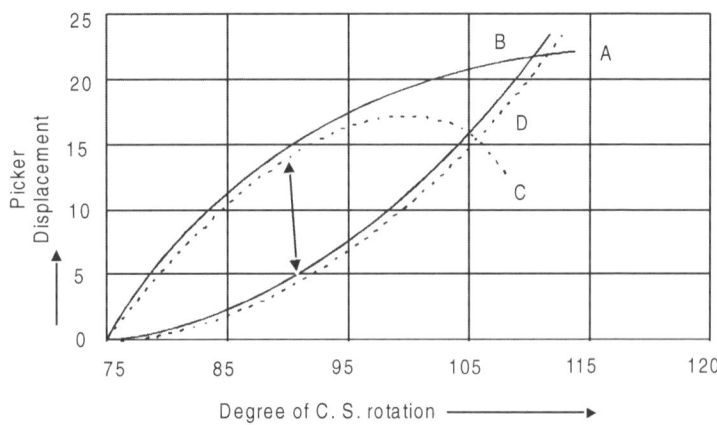

13.1 Nominal and actual displacement of picker.

$$v^2 = 2 \text{ as, and } f = ma,$$

We get,

$$f = (mv^2)/2s.$$

where,

v = the final velocity (m/s),
a = the uniform acceleration (m/2),
f = force (N),
s = the distance over which acceleration occurs (m),
m = mass of the shuttle (kg).

Putting the values in the above equation, we obtain

$$f = (0.5 \times 12.52)/2 \times 0.2 = 196 \text{ N}.$$

Now, comparing the above result with the one obtained for nominal displacement to overcome friction, we find that the force required to overcome the inertial force is about six times the force required to overcome friction. It will be seen later that the acceleration is never uniform, and the peak force generated in overcoming picking may be as much as twice what would be if the acceleration were uniform.

For our present discussion, the force required to overcome the friction is neglected. This is justified in the sense that it is relatively small compared with the force required to overcome the inertial force.

Going back to Fig. 13.1, the curves (A) and (B) are due to lag produced by the force required to overcome inertia of the shuttle. The picking mechanism is not rigid, and the stresses set up in the mechanism by the resistance o shuttle to its acceleration result in strains in the picking mechanisms: the picking bend or lug strap stretches, the picking stick bends, and the picking shaft twists. In the early of the shuttle acceleration (0° to 15°), these stresses and strains are building up because the picker is trying to overcome the shuttle resistance. In the later part (15° to 30°), the stresses and strains diminish and ultimately disappear when the shuttle leaves the picker where the two curves.

The action of the picker, the picking band and the picking stick can be compared to that of elastic of a catapult and the shuttle as the missile – this analogy has been forwarded by Thomas. The maximum lag at about 15° can be compared to the straining of the catapult, and the second half of 15° to 30° can be compared to the release of such elastic strain. As soon as the elastic becomes slack the missile, that is the shuttle in our case, leaves the leather, i.e., the picker, at the intersection of the two curves.

If there is a linear relation between stress and strain during picking, the lag at any instant would be proportional to the force exerted on the shuttle by the picker at that instant. Because of other complicating factor this is approximately true. Even if it is accepted as approximation, it is clear that the force of acceleration increases from zero at the commencement of a pick to a maximum about halfway through the pick and then falls to zero as soon as the shuttle leaves the picker. If it were uniform the peak value would be larger than that would be needed.

The two additional curves (C) and (D), shown in Fig. 13.1, are for cone underpick in automatic weaving machine. In absence of picking band and the type of picking stick used in overpick machine, the picking mechanism is more rigid. In this case the acceleration is complete over a short distance but over the same angular movement of the crankshaft. The width of the loop at 15° is less than the one between the curves (A) and (B), which shows that the lag in this case is less, but none the less the picking stick does have a tendency to bend under the influence of shuttle's inertia force.

The use of picking tappet designed to give uniform nominal acceleration, does not give uniform actual acceleration, but gives fluctuating acceleration with a peak value equal to twice the designed value. Repeated applications of such a transient force have several undesirable consequences. Lateral vibration of large amplitude is set up in the machine frame; there is excessive wear of the teeth of the two main gear wheels in mesh during acceleration, resultant in loss of machine speed; the checking of picking mechanism, which has to be made sufficiently robust to withstand oscillating stresses imposed on it, leads to wasted energy and more vibration.

The maximum force involved in picking should be reduced, and it can he achieved in two ways: by accelerating the shuttle more uniformly and accelerating it over a longer distance. The best possible conditions would be obtained by accelerating the shuttle uniformly over the whole available length of the shuttle box. Improvement obtained by increasing the distance without increasing the degree of uniformity of acceleration was demonstrated by Thomas and Vincent by use of double length of rubber tubing. This showed a reduction in force of about 40% compared to one when normal picking band is used. But as the rubber tubing of sufficient durability having large extension is not industrially available it became an academic interest only.

13.3   Elastic properties of picking mechanism

Then the stiffness of the a picking mechanism is known and it is easy to measure under static conditions, its use in conjunction with the nominal and actual displacement curves to estimate the force exerted by the picker on the shuttle during picking and to compare these with constant force that would produce the same final shuttle speed. Thomas and Vincent obtained certain figures on which the following calculation are based and are given in Table 13.1, and relates to loose-reed over pick machine,

Table 13.1

Mass of shuttle	0.32 kg
Stiffness of mechanism	3.65 kN/m
Shuttle speed	12.2 m/s
Stroke of picker	0.20 m
Maximum Lag	0.075 m

Since a force of 3.65 kN acting on the picker parallel to the shuttle axis produces a deflection of 0.075 m in force required to produce such lag is

$$0.075 \times 3650 \ N = 274 \ N,$$

This is the peak force exerted by the picker on the shuttle during picking. Now, the uniform force that would produce the same final speed over the same distance is given by

$$f = (mV^2) \ / \ 2S = (0.32 \times 12.2^2) \ / \ (2 \times 0.2) = 119N$$

The actual peak force of 274N includes the force required to overcome friction. A typical values of 32.5N for this, so it may be reasonably assumed that, of actual peak force of 270N, a force of $(274 - 32.5) \ N = 241.4N$ was used to overcome inertia. This is about twice the uniform force of 119N.

Forces encountered in the modern automatic weaving machines are substantially greater than these calculated above because the shuttle tends to be heavier, the effective stroke shorter, and the speed of the shuttle greater. In the machine to which the curves (C) and (D) in Fig. 13.1 relates the shuttle speed was 13.3 m/s, the mass of the shuttle was 0.51 kg including a fill pirn, and the effective length of stroke was 0.165 m. The required uniform accelerating force would be

$$f = (0.51 \times 13.3^2) \ / \ (2 \times 0.165) = 273.5N$$

This compares with 119N in the proceeding example, in which the actual peak force was twice as great. The actual peak force in the last example might exceed 500N.

The above calculations assumed the following conditions:

i. The picking mechanism as a simple elastic system obeying Hooke's law,

ii. The stiffness of the system remained constant the picking,

iii. The force acting on the shuttle given by the product of the stiffness and lag remained constant.

But these assumptions are not completely true.

The stiffness of the picking system close to the main drive is different from that of from the side opposite to the main drive, this is attributed to the twisting of the camshaft which results in reduced stiffness during picking from the other side, and is more true for wider machines. Lord has shown that the stiffness may substantially increase from the beginning to the end of a pick.

Thus, that picking mechanism is analogous to a simple elastic system is only of approximate nature.

13.4   Initial and average shuttle speed during traverse

Knowledge of the relation between the two speeds is desirable in order to co-ordinate the picking with shedding. It is clear, however, that if the shuttle is pressed between the two sheets of warp during the part or its entire traverse, no definite relation between the two speeds can exist. Thomas and Vincent did sonic experiment to find the difference and observed that there was only slight difference and it could be easily ignored unless some critical value affecting other factors come into play. Some typical values are given below in Table 13.2.

Table 13.2

Crank position(after beat-up)	Initial speed (m/s)	Final speed (m/s)
60θ	10.4	10.1
75θ	10.7	10.5
120θ	11.5	11.2

They found that the initial and average shuttle speeds are approximately equal and ranged from 9.4 to 14.6 m/s. The sizes of the nose bits of over pick system are approximately directly proportional to the nominal displacement after 30° of the crank-shaft rotation.

The shuttle speed is approximately directly proportional to the machine speed and inversely proportional to the distance from the picking shaft. Thus we write

$$vs - v/d$$

where,   v = machine speed
         d = distance picking tappet to tappet shaft

They also found that the shuttle speed is independent of the shuttle mass and buffer position, shortening of picking band, lowering of the picking bowl increases the shuttle speed. When picking time is progressively made later, the shuttle speed is seriously lowered due to the greater interference in shed, i.e., the bending factor is reduced, as the shuttle leaves the shed. With greater reed width the relative shuttle speed, which is defined as the ratio between the average shuttle speed and the machine speed, is also increased.

13.5   Factors affecting initial shuttle speed

There are several factors which affect the initial shuttle speed in an over pick machine which were thoroughly investigated by Thomas and Vincent. The primary factors which are responsible for controlling the shuttle speed are discussed below:

13.5.1 Shape of picking tappet

Shape of that part of the picking tappet in contact with the picking bowl, while the shuttle is being accelerated, other things being unaltered, it will depend on the length of the nose bit of the picking tappet. Depending on the size of the machine width, the nosebits are marked by punches to indicate the range of shuttle speed that would be required for a machine. Wider the machine greater the number of punches marked.

13.5.2 Machine speed

The average machine speed during the shuttle traverse does not matter materially, that is, during picking the machine speed does not differ much. But there is evidence at present time that, using sophisticated measuring instruments, machine speed does fall during shuttle acceleration. However, it will he evident from the figures below that with increasing machine speed the average shuttle speed also increases but not strictly in direct proportion.

| Machine speed (ppm) : | 216 | 202 | 165 | 171 |
| Shuttle speed (m/c)   : | 13.0 | 12.7 | 119 | 11.7 |

13.5.3 Time of picking

Variation in the time of picking has a tendency to give higher shuttle speed when picking was formed to be late with respect to the crankshaft position starting from 45°. The changes in speed with timing were less with under pick when with over pick system, and were absent with over pick when the sley was fixed. Higher shuttle speed is necessary as it has to force through the shed as it progressively lowers as the shuttle emerges out of shed.

13.5.4 Length of picking band

Changing the length of the picking band is one of the simplest and most frequently adjusted part of the picking mechanism that is handled by the tacklers or weavers during routine weaving. It is also a change that may occur spontaneously through stretching, if the band is new, through recovery from stretch when the machine is stopped, and through changes in the humidity. Such changes produce double effects, it alters the time of picking and also the position of the picking bowl on the nosebit at the instant the hand is taut. This also results in change of picking force. The length of picking band has a direct bearing on the shuttle speed – greater the band length lesser the speech and vice-versa.

13.5.5 Swell resistance

The swell exerts a frictional retardation force on the shuttle and before the start of the shuttle it is constant. The shuttle will not move until the elastic displacement of the picking mechanism gives rise to a force on the shuttle sufficient to overcome this frictional force due to the swell pressure. Changes in the swell resistance may affect the shuttle speed in two ways. There is a direct effect resulting from changes in the resistance offered to the shuttle during acceleration and there may be an indirect effect if the swell pressure is great enough resulting from variation in the position at which the shuttle comes to rest after the previous pick, since variation of the starting position modifies the nominal movement during picking. With increased swell resistance the shuttle speed is increased due to straining of the picking mechanism. With the sudden release of the swell resistance, an unbalanced force is available for shuttle acceleration. Some of the modern under pick weaving machine have in them incorporated the swell releasing or "easing" motion, and this precludes any unbalanced force being released during picking. The effect of swell resistance is shown against the shuttle speed below:

Max, swell resistance (N)	:	2.24	8.95	20.01	27.60	32.10
Shuttle speed (m/s)	:	12.4	12.6	12.7	12.75	12.90

13.5.6 Mass of shuttle

The amount of checking, a shuttle requires, is dependent on its mass and its speed. Reduction in momentum is the direct result in reduction of mass and not its speed. It is often viewed that a lighter shuttle moves more slowly than a heavier one – but this is not true. From experimental observations, it was found that the shuttle speed is substantially independent of its mass. Timing was not affected though the heavier shuttle has a tendency to be picked later.

13.5.7 Height of the picking bowl

It was observed that within the limits set by machine designers raising of the picking bowl causes the reduction in shuttle speed, and the following figures substantiate the facts:

Height of picking bowl axis

above the floor (cm) :	40.70	41.30	41.30	42.70
Shuttle speed (m/s) :	10.30	10.20	9.40	9.20

13.5.8 Picking tappet distance from picking shaft

For a given increment in the radius of the picking tappet at its point of contact with its picking bowl, there will be a large increase in the angular movement of the picking shaft where the tappet is nearer to it than when the picking shaft is further away. For this reason an increased shuttle speed would be expected as the picking tappet was moved nearer to the picking shaft. The relation between this parameter and the shuttle speed is given below:

Distance between the central plane of picking tappet and

axis of picking shaft (cm) :	10.0	11.3	12.5	13.5
Shuttle speed (m/s) :	11.9	11.2	10.2	9.4

13.5.9 Position of buffer

It is viewed that if the picking mechanism is checked immediately after the picking tappet has ceased to exercise control, the "pick will become chocked" and the shuttle speed will be lowered. This contention has been negative by experiments. Progressively increasing the distance between the spindle stud and buffer showed that the shuttle speed remained unaltered. Thus under the circumstances the position of buffer has no bearing on the shuttle speed.

13.5.10 Prolonged weaving

It is desirable that there should be no changes in shuttle speed as weaving continues. Such changes may occur through the working of loose nuts and bolts, changes in the length of the nicking band, and in the changes in the degree of checking. Whilst the first and the last named causes are not systematic in nature and hence not useful, the picking band stretches and causes lowering of the shuttle speed. Though this stretching is not a continuous process, as bands recover when the machine is stopped, resulting in increased speed on restarting. After sometime stretching ceases and changes in shuttle speed is also stopped. With well regulated relative humidity and temperature, the effects of other disturbances are small.

13.5.11 Initial gap between picker and shuttle

From time to time in course of routine, weaving the shuttle may be obstructed in its passage through the shed and causes a gap between the shuttle and the picker. This naturally leads to a reduction in shuttle speed for the next pick, causing "loom bang-off" or shuttle trap. With ordinary nose-bits the fall in speed becomes serious when the initial gap exceeds 25 mm, but with constant nominal acceleration the fall is not that serious.

13.6   Shuttle checking

After the shuttle leaves the picker at the end of picking it enjoys a free flight across the shed until it reaches the other box. However, during this free flight there is some retardation due to

1. friction with the bottom warp shed line, and
2. friction between the shuttle edge and the reed.

The usual retardation of a shuttle for a slow running machine is in the vicinity of 6.1 m/s^2, and it is about 10.7 m/s^2 in case of a fast running machine.

From the equations of motions we find that the maximum velocity of shuttle as it enters the shed by

$$v = (s + 0.5 \, at^2)/t$$

where,   s = total shuttle flight (m),
          a = acceleration (m/s^2)
          t = flight time (s),

If the width of the machine plus the effective length of the shuttle is 1.08 m and the machine speed is 190 ppm, the shuttle retardation is 10.7 m/s^2, and if the shuttle takes 105° of the crankshaft rotation, the time for the shuttle flight is

$$t = (105 \times 60) / (360 \times 190)$$
$$= 0.092 \text{ s}$$

Then the maximum shuttle velocity is found out from the above relation:

$$v = (1.08 + 0.5 \times 10.7 \times 0.092^2) / 0.092$$
$$= 12.23 \text{ m/s}$$

It can be seen from Fig. 13.1, that after about a width of 200 cm, the maximum shuttle velocity changes very slowly in increase in the machine width so that it can be ignored.

Velocity of the shuttle during traverse is found if we assume that at any instant, say, t_x, during the total time of traverse t, let the shuttle velocity be v_t, then

$$v_t = v_{-atx}$$

Also therefore, the speed of the shuttle at the end of its flight or on emerging out of the shed at time t,

$$v_t = v_{-at}$$

where,

$$v_t = \text{terminal velocity (m/s)}$$

Therefore, if the maximum shuttle velocity be 1.23 m/s and the shuttle retardation is 10.7 m/s^2 and its traverse time is 0.092 s, then the terminal velocity of shuttle at the end of its flight would be

$$V_t = 12.23 - 10.7 \times 0.092$$
$$= 11.27 \text{m/s}$$

So, it is found that the shuttle emerges out or the shed at a slightly lesser speed that it was projected into the shed, and it must be brought to rest at a reasonably smoother way over a distance about the same as, or ratter less than that over which it was accelerated. This become increasingly difficult as the speed of the machine increases, since the energy that has to be dissipated, that is the kinetic energy of the shuttle, is proportional to the square of it velocity. As because the problem of shuttle checking is one of the main factors that limit the speed of the machine, the design of high speed machine must be accompanied by very efficient checking system.

Shortly after the leading end enters the shuttle box, the shuttle strikes the swell which is usually situated at the back of the shuttle box, though in some makes it is found in front of the box as well, the swell is displaced.

The swell has two principle functions:

i. It operates the warp protection device in a fast reed machine and monitors the shuttle movement, that is, it acts as a detecting system, and prevents any damage to warp;

ii.   It helps to reduce the shuttle speed at the end of it trajectory.

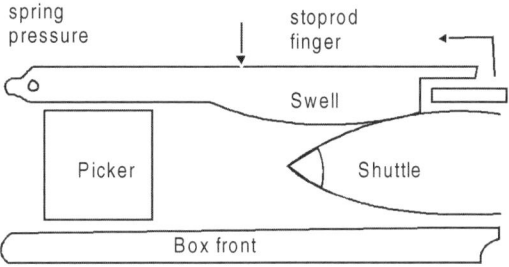

13.2 A simple swell.

Figure 13.2 shows a simple arrangement of a shuttle checking device in a fast reed weaving machine. In this system the swell is pivoted at the outside of shuttle box and is spring loaded either by means of a flat or spiral, depending on the on the machine design, and protrudes inside the box.

As the shuttle strikes the swell, it is displaced, thus causing the stop-rod finger to be displaced at the same instance. The stop-rod finger is also spring loaded. Pressure due to the springs cause the shuttle to slow down as the swell assembly resists any displacement. The pressure between the swell and shuttle, and between shuttle and box front are caused by the swell spring or the stop = rod spring or both. In some cases additional pressures are called into play which is released during picking to avoid excessive swell pressure. This mechanism of easing the swell pressure is often known as the swell releasing mechanism.

The shuttle retardation curve obtained by Thomas and Vincent for a fast reed machine is shown in Fig. 13.3.

Where it is found the shuttle strikes the swell at about 14.3 m/s, it is found that during the next 0.114 m of travel of the shuttle its velocity is reduced by about 32%, i.e., the velocity becomes 9.8 m/s when the striking picker (B). Thereafter the combined action of the picker and swell and the cheek strap reduces the shuttle velocity to zero over a further distance of 0.05 m of it travel. It is clear that the retardation is not uniform. The actual retardation between the shuttle striking the swell first and its contact with the picker at (B) is found by using the relation:

$$V^2 = U^2 - 2as,$$

and putting the values from the curve, we have

$$9.8^2 = 143^2 - 2a \times 0.114$$

or, $a = 478$ m/s^2

Again, the shuttle is brought under the action of the checking system (check strap) at the point C, and during this period an actual retardation of about 2000 m/s^2 takes place.

The dotted line in the figure shows the uniform retardation curve over the total distance of 0.165 m. Under this condition the uniform retardation between the first contact of shuttle with the swell and the position of (B) would be 616 m/s^2. Thus it can be said that the swell failed to reduce the shuttle speed, and this had been verified in the actual observation that the pivot type swell, on contact with the incoming shuttle flies back, and there is a momentary lose of contact between the two, and during this period the shuttle moves at a fairly high speed inside the box. This type of contact and slapping takes place for a considerable length of time until the shuttle comes to rest. Thus the curve takes the form of slip-stick nature. The curve further reveals that at the point of contact with the picker the shuttle is retarded heavily with considerable reduction in shuttle speed—the actual curve takes a dip under the dotted curve at (C). Thus throughout the retardation period inside the shuttle box the shuttle is never retarded uniformly, which is due to sley construction mainly.

More efficient checking can be obtained by separating the two functions of the swell. In the modern weaving machines the tendency is to use two swells or a swell in two parts. Attempts were made by Roy with the use of "piano swell" where as many as eight mini swells, each being individually spring loaded, in the form of piano keys, were used. Though the condition of uniform retardation was quite appreciably good, it was thought that to maintain such a number would be a problem in actual practice. Roy gradually reduced the number of such swells to two through four, and the results were very encouraging.

Figure 13.4 shows one of the arrangements of swell in a modern machine and the shuttle retardation is illustrated in Fig. 13.5. In this type of arrangement it is found that a small swell at the box mouth is hinged, and this is connected to the stop rod system. This part thus functions as the shuttle monitoring system. This small swell has a very small effect, on the shuttle retardation. The second swell, which floats on two spiral springs, is the main swell and it is to be noted that this one is not hinged like the former type. As the shuttle enters the box the front end of the swell is pressed back at position (B) of Fig. 13.5. This will cause the rear end of the swell to retract a little, but never will fly back as in the pivoted case. Both the ends of the swell retract at position (c), Fig. 13.5. From this

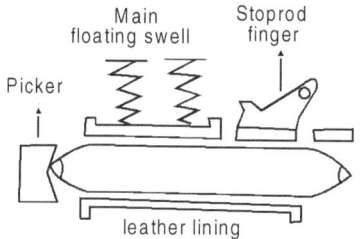

13.4 Floating swell (cm).

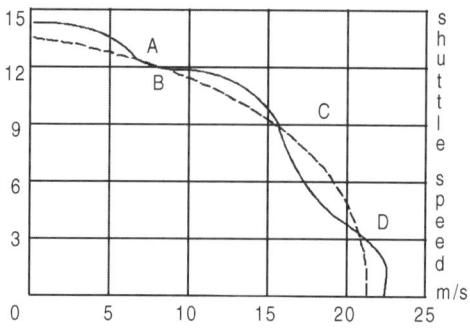

Dist. moved by shuttle after striking the swell

13.5 Shuttle retardation curve (floating swell).

position onwards the shuttle speed is reduced fairly uniformly and rapidly. As the shuttle strikes the picker at (D), the check strap cushions the speed. In the present machines this cushioning is done by hydraulic system instead of check strap.

If we compare the two curves we find that the curves at Fig. 13.5 has a lesser gradient than that of the curves in Fig. 13.3, which indicates that deceleration in the second case is more uniform and takes place dyer a longer distance. This shows a substantial improvement in shuttle checking. In place of using wooden swells, two metal units with greater moment of inertia provide progressive action on the shuttle so that it is accurately positioned in the box. Swells with greater moment of inertia have been found to absorb the kinetic energy of the shuttle in a shuttle box more efficiently.

13.6.1 Shuttle mass and checking

The mass of the shuttle gradually decreases as the pirn weaves down. However, the checking force remains unaltered, and hence the effectiveness of checking the shuttle must have to be better. As the shuttle mass is gradually reduced the impact velocity of the shuttle will also be less. The checking system should be efficient to allow for variation in the shuttle speed due to reduction of its content. Furthermore, it should be able to allow for any variation in the shuttle speed due to friction and other resistances during its trajectory through the shed.

Consider, for example, the mass of the shuttle including its contents to be 0.51 kg when the pirn is full, and it becomes 0.48 kg when it is nearly empty, that is during weft insertion 30 g of yarn have been consumed. Let us also assume that the impact velocity of the shuttle should not be less than 4.5 m/s at any time during correct functioning of the machine, its calculated speed is 13.75 m/s as the shuttle strikes the swell, and is uniformly retarded over a distance of 0.20 m up to impact with the picker.

Now,

$$V^2 = u^2 - 2as$$

where,

U = initial velocity,
V = impact velocity,
a = uniform retardation,
s = retardation distance

The impact velocity is least when the pirn is nearly empty, i.e., when

$$13.75^2 - 4.5^2 = 2a \times 0.20$$

thus,
$$a = 422 \text{ m/s}^2.$$

Since, f = ma, retardation is inversely proportional to the ass being retarded, so that when the shuttle is full, we have

$$a = 422 (0.48 \times 0.51) = 397 \text{ m/s}^2$$

and
$$13.75^2 - V^2 = 2 \times 397 \times 0.20$$
$$V = 5.5 \text{ m/s}$$

In this example, the impact velocity is about 22% greater than the pint when it is full than when it is nearly empty.

13.6.2 Ideal checking conditions

It is not difficult to lay down conditions which checking should fulfill. They are

i. the shuttle should come to rest in contact with the picker at the same place in the shuttle box after each picks
ii. the maximum value of retardation should be kept as small as possible, and
iii. the impact velocity of the shuttle with the picker should be as small as possible.

These conditions are entirely independent of each other, since (iii) is largely implied by (ii), for impact always results in rapid deceleration. The first condition aims at ensuring that the shuttle speed is uniform in the following pick; the second aims at reducing to a minimum the force tending to displace and disintegrate the weft package; and the third condition aims at reducing the wear rate of the picker.

13.6.3 Checking limits

The checking limit is determined by the ability of the checking system to absorb the kinetic energy of the shuttle in the time available between two successive arrivals in any one box. As the shuttle is brought to rest, its kinetic energy is transformed into heat and the checking limit is therefore essentially dependent on the heat-disposal capabilities of the checking system. If the final velocity of the shuttle is too high, the shuttle itself, the swell and other parts of the box mechanism will become hotter and hotter, and this will eventually cause damage that will make weaving impossible. The heat disposal capabilities of the checking system can be expressed by the amount of heat in units of mechanical energy that can be dissipated per second. As the checking limit is intimately connected with the picking limit, they should be kept in harmony. Where the picking limit, i.e. the maximum initial velocity of the shuttle, is constant, the full benefits are only realized when the heat disposal potential of the checking system is high enough to bring the whole range of weaving machines within the orbit of this limit. With the lowest value of heat disposal capability, the speed potential of the picking motion is entirely wasted because the whole range is governed by the checking limit. Thus, any raising the rising picking limit would benefit the wider machines. It is desirable that in order to increase the picking limit, the design and material used in the checking system and the shuttle itself, much careful attention must be given.

Ishida et al during their work have measured the temperature in the checking system which are summarized below and are of considerable interest to the designers as well as those engaged with cloth production:

(a)   the maximum temperature of the swell caused by friction with the shuttle was about 65°C and was at the crest of the smell profile,

(b)   the temperature of the swell was affected by the kinetic energy of the shuttle, the maximum being at the same point as that of higher retardation of the shuttle,

(c)   the maximum temperature of the swell was reached about 20 min after the machine started, and

(d)   the temperature of the swell at a machine speed of 300 pm would be reasonable to maintain the quality and effectiveness of the swell cover used.

They worked on a 135 cm reed space automatic single shuttle V machine using a shuttle weighing, with pirn, 435 g. The above findings are based on machine running at 154 ppm. Such raising of temperature causes charring of the shuttle back which could be seen clearly after certain length of time.

PART II

Calculations of textile machinery

Fundamentals of motion in textile machinery

14.1   Introduction

The chapter deals with the following aspects:

(i)   *Kinematics and kinetics of motion*, i.e. the relative motion of bodies with and without consideration of the forces causing the motion. Kinematics deal with the geometry of motion and concepts like displacement, velocity and acceleration considered as functions of time. Kinetics of motion deals with the motion which takes into consideration the forces or other factors, e.g. mass or weight of bodies. The force and motion is governed by the three laws of motion.

(ii) *Friction* is the resistance offered by one body to another during motion, due to the irregularities present in the surfaces of the bodies. It plays a vital role in a number of textile processes like spinning, winding and weaving. It is necessary to reduce the friction as otherwise abrasion and wear would result.

(iii) *Various methods of drives* such as belt and rope drives and gear drives. The belt and rope drives are negative methods of power transmission. Whereas the gear drives are positive methods of transmission. In the negative method of power transmission, there is loss due to slippage, which is not found in gear drives.

14.2   Basic definitions

Plane motion

When the motion of a body is confined to only one plane, the motion is said to be plane motion. The plane motion may be either rectilinear or curvilinear.

Rectilinear motion

It is the simplest type of motion and is along a straight line path. Such a motion is also known as translatory motion.

Curvilinear motion

It is the motion along a curved path. Such a motion, when confined to one plane, is called plane curvilinear motion.

Linear displacement

It may be defined as the distance moved by a body with respect to a certain fixed point. The displacement may be along a straight path or a curved path. The displacement of body of a body is a vector quantity, as it has both magnitude and direction. Linear displacement may, therefore, be represented graphically by a straight line.

Linear velocity

It may be defined as the rate of change of linear displacement of a body with respect t time. Since velocity is always expressed in a particular direction, therefore it is a vector quantity. Mathematically, linear velocity, v, is expressed as

$$v = ds/dt$$

where s is displacement and t is time. If the displacement is along a circular path, then the direction of linear velocity at any instant is along the tangent at that point. The speed is the rate of change of linear displacement of a body with respect to the time. Since the speed is irrespective of its direction, therefore it is a scalar quantity.

Linear acceleration

It may be defined as the change of linear velocity of a body with respect to the time. It is also a vector quantity. Mathematically, linear acceleration, a, is expressed as

$$a = \frac{dv}{dt} = \frac{d\left(ds/dt\right)}{dt} = \frac{d^2s}{dt^2} \text{ (because } v = ds/dt)$$

The linear acceleration may also be expressed as follows

$$a = \frac{dv}{dt} = \frac{ds}{dt} \times \frac{dv}{ds} = v \times \frac{dv}{ds}$$

The negative acceleration is known as deceleration or retardation.

Angular displacement

It may be defined as the angle described by a particle from one point to another, with respect to the time. It is a vector quantity since it has both magnitude and direction.

Angular velocity

It may be defined as the rate of change of angular displacement *(θ)* with respect to time. It is usually expressed by a Greek letter ω (omega). It is also a vector quantity. Mathematically, angular velocity is expressed as

$$\omega = \frac{d \cdot \theta}{dt}$$

Angular acceleration

It may be defined as the rate of change of angular velocity with respect to time. It is usually expressed by a Greek letter α (alpha). Mathematically, angular acceleration is expressed as

$$\alpha = d\omega/dt = \frac{d}{dt}(d.\theta / dt) = \frac{d^2\theta}{dt^2} \quad [\text{as } \omega = d.\theta/dt]$$

It is also a vector quantity, but its direction may not be same as that of angular displacement and angular velocity.

Mass and weight

$$a = \frac{d\omega}{dt} = \frac{d}{dt} \times \frac{d \cdot \theta}{dt} = \frac{d^2 \cdot \theta}{dt^2} \quad [\text{as} \omega = d.\theta / dt]$$

Mass is the amount of matter contained in a given body, and does not vary with the change in its position on the earth's surface. The mass of a body is measured by direct comparison with a standard mass by using a lever balance.

Weight is the amount of pull, which the earth exerts upon a given body. Since the pull varies with the distance of the body from the centre of the earth, therefore the weight of the body will vary with its position on the earth's surface. Thus, it is a force.

Momentum

It is the total motion possessed by a body. Momentum = mass × velocity

Let   *m* = mass of the body,

        u = initial velocity of the body,

        v = *final* velocity of the body,

        a = constant acceleration, and

        t = time required (in seconds) to change the velocity from *u* to *v*.

Now, initial momentum = $m \times u$
and final momentum = $m \times v$
Therefore change of momentum = $m \times v - m \times u$

And rate of change of momentum = $\dfrac{m \times v - m \times u}{t} = \dfrac{m(v-u)}{t} = m.a$

Force

It may be defined as an agent that produces or tends to produce, destroys or tends to destroy motion.

$$\text{Force} = \text{mass} \times \text{acceleration}$$

Moment of a force

It is the turning effect produced by a force on the body on which it acts. The moment of a force is equal to the product of the force and the perpendicular distance of the point about which the moment is required, and the line of action of the force.

$$\text{Moment of a force} = F \times l$$

where F = force acting on the body, and
l = perpendicular distance of the point and the line of action of the force.

Couple

A couple is the effect of two equal and opposite parallel forces, whose lines of action are different. If F is the force and X is perpendicular distance between their lines of action then

$$\text{Moment of a couple} = F \times X$$

Centripetal and centrifugal force

Consider a particle of mass m moving with a linear velocity v in a circular path of radius r. The centripetal acceleration, a_c is expressed as

$$a_c = v^2/r = \omega^2. r$$

We know that, Force = Mass × acceleration
Therefore, centripetal force *(Fc)* = mass × centripetal acceleration
or $Fc = m.v^2/r = m \bullet \omega^2. r$

This force acts radially inwards and is essential for circular motion. The centripetal force acts radially inwards. According to Newton's third law of motion, action and reaction are equal and opposite. Therefore, the particle must exert a force radially outwards of equal magnitude. This force is known as centrifugal force whose magnitude is given by

$$Fc = m \bullet v^2/r = m. \omega^2. r$$

Centripetal and centrifugal forces are of consideration in the motion of ring and traveler in a ring frame.

Mass moment of inertia

It is the sum of the mass of every particle of a body multiplied by the square of its perpendicular distance from a fixed line.

The moment of inertia of a body is given by $I = m.k^2$

where, I – Moment of inertia,

m – Mass of the body, and

k – Radius of gyration.

The unit of moment of inertia in SI unit is kg m².

The moment of inertia is an important consideration in the loom slay, which governs vibration in a loom due to beat up.

Note:

1. If the moment of inertia of a body about an axis through its centre of gravity is known, then the moment of inertia about any other parallel axis may be obtained by parallel axis theorem, i.e. moment of inertia (Ip) about a parallel axis,

$$Ip = I_G + m. h^2$$

Where, I_G = moment of inertia of a body about an axis through its centre of gravity, and

h = distance between the two parallel axes

2. The following are the values of I for simple cases:

   (a) The moment of inertia of a thin disc of radius r, about an axis through its centre of gravity and perpendicular to the plane of the disc is

   $$I = m.r^2/2$$

   and, moment of inertia about a diameter,

   $$I = m \bullet r^2/4$$

   (b) The moment of inertia of a thin rod of length l, about an axis through its centre of gravity and perpendicular to its length,

   $$I_G = m. l^2/12$$

3. The moment of inertia of a solid cylinder of radius r and length l, about the longitudinal axis or polar axis is

$$m.r^2/2$$

and moment of inertia through its centre perpendicular to longitudinal axis is $r^2/4 + l^2/12$

Radius of Gyration

It is the distance, from a given reference, where the whole mass of body is assumed to be concentrated to give the same value of mass moment of inertia.

Angular momentum or Moment of momentum

It is defined as the product of mass moment of inertia and the angular velocity of the body.

Torque

It may be defined as the product of force and the perpendicular distance of its line of action from the given point or axis.

Work

Whenever a force acts on a body and the body undergoes a displacement in the direction of the force, then work is said to be done.

Work done = Force × displacement
Work done = Torque × angular displacement

The unit of work depends upon the unit of force and displacement
In the SI system of units, the practical unit of work is N-m.

Power

It may be defined as the rate of doing work or work done per unit time.

$$\text{Power} = \frac{\text{work done}}{\text{time taken}}$$

Note:

1. If T is the torque transmitted in N-m or J and ω is the angular speed in rad/s, then

Power, $P = T. \omega = T \times 2\pi N/60$ watts $(\omega = 2\pi N/60)$
Where N is the speed in rpm
2. The ratio of power output to power input is known as efficiency of a machine. It is always less than unity and is represented as percentage

$$\text{Efficiency, } \eta = \frac{\text{power output}}{\text{power input}}$$

Energy

It may be defined as the capacity to do work. The energy exists in many forms e.g. mechanical, electrical, chemical, heat, light, etc.

The mechanical energy is equal to the work done on a body in altering either its position or its velocity. The following three types of mechanical energies are important:

(a) Potential energy

It is the energy possessed by a body for doing work, by virtue of its position. For example, a body raised to some height above the ground level possesses potential energy because it can do some work by falling on earth's surface.

The potential energy *P.E.* is given by,

$$\text{P.E.} = W.h = m.g.h \qquad (W = m.g)$$

Where W = weight of the body,
m = mass of the body,
h = distance through which the body falls, and
g = acceleration due to gravity.

It may be noted that when W is in Newton and h in meters, then the potential energy will be in N-m.

When m is in kg and h in meters, then the potential energy will also be in N-m.

(b) Strain energy

It is the potential energy stored by an elastic body when deformed. A compressed spring possesses this type of energy, because it can do some work in recovering its original shape.

The strain energy is given by,

Strain energy = work done $\frac{1}{2}.W. n = \frac{1}{2}.s .n^2$ $(W = s \times n)$

Where, s = stiffness of compressed spring,
n = distance through which spring is deformed, and
W = load

In case of a torsional spring of stiffness q N-m per unit angular deformation when twisted through an angle θ radians, then

$$\text{Strain energy} = \text{work done} = \tfrac{1}{2}.q.\theta^2$$

(c) Kinetic energy

It is the energy possessed by a body, for doing work, by virtue of its mass and velocity of motion. If a body of mass m attains a velocity v from rest in time t, under the influence of a force F and moves a distance s, then

$$\text{Work done} = F.s = m.a.s$$

Therefore kinetic energy of the body or the kinetic energy of translation *(K.E.)*,

$$K.E. = m.a.s = m \times a \times v^2 = \tfrac{1}{2}\,m.v^2$$

We know that, $v^2 - u^2 = 2a\,s$

Since $u = 0$ because the body starts from rest, therefore

$$v^2 = 2a.s \text{ or } s = v^2/2a$$

Note:

1. When a body of mass moment of inertia I (about a given axis) is rotated about the axis, with an angular velocity ω, then it possesses some kinetic energy. In this case,

$$\text{Kinetic energy of rotation} = \tfrac{1}{2}\,I.\omega^2$$

2. When a body has both linear and angular motions, for e.g. in the locomotive driving wheels and wheels of a moving car, then the total kinetic energy of the body is equal to the sum of the kinetic energies of translation and rotation.
   Therefore total kinetic energy $= \tfrac{1}{2}.m.v^2 + \tfrac{1}{2}I.\omega^2$

14.3   Equations of motion

Practical importance

The equation of linear motion is concerned with motion in a straight line and may be applied in such cases as finding the velocity of the shuttle in a loom.

The equation of angular motion is concerned with angular or rotational movements. Ideal cases of application can be the rotation of the crank shaft in a loom.

14.3.1 Equation of linear motion

1. $v = u + a \cdot t$
2. $s = u \cdot t + \frac{1}{2} a \cdot t^2$
3. $v^2 = u^2 + 2a\,s$

4. $s = \dfrac{(u + v)}{2} \times t$

       where u = initial velocity of the body
           v = final velocity of the body
           a = acceleration of the body
           s = displacement of the body in time t, and
           v_{av} = average velocity of the body during the motion.

Note:

1. The above equations apply for uniform acceleration. If, however, the acceleration is variable, then it must be expressed as a function of either t, s or v and then integrated.
2. In case of vertical motion, the body is subjected to gravity. Thus g (acceleration due to gravity) should be substituted for 'a' in the above equations.
3. The value of g is taken as +9.81 m/s^2 for downward motion and –9.81m/s^2 for upward motion of a body.
4. When a body falls freely from a height h, then its velocity v, with which it will hit the ground is given by $v = \sqrt{2gh}$

14.3.2 Equations of angular motion

1. $\omega = \omega_o + \alpha \cdot t$
2. $\theta = \omega_o.t + \frac{1}{2} \alpha .t^2$
3. $\omega^2 = (\omega_o)^2 + 2\alpha \cdot \theta$

4. $\theta = \dfrac{\omega + \omega_o}{2}$

       where ω_o = initial angular velocity in radians/s,
           ω = final angular velocity in radians/s,
           t = time in seconds,
           θ = angular displacement in time I seconds, and
           α = angular acceleration in radians/s^2

Note: If a body is rotating at the rate of N rpm, then its angular velocity,

$$\omega = 2\pi N/60 \text{ radians/s}$$

The relation between the mass (m) and weight (w) of a body is $w = m \cdot g$ or $m = w/g$

where w is in Newton, m is in kg and g is the acceleration due to gravity in m/s²

14.4   Vibratory motions

There are two types of motions that occur frequently in textile operations, namely the linear and circular types. There is a third type of motion that connects both the motions and is known as the simple harmonic motion (SHM). The movement of a pendulum bob or of a tuning fork serves as a good example. The SHM can be of two types, namely linear and angular. In the first type the circular motion can be converted into reciprocating motion and vice versa. Examples of the former include the movement of a thread guide bar in reeling and the shedding griffe in a jacquard loom. The angular SHM is encountered in the vibration of a simple pendulum.

The following are the types of vibrations:

(a)   Damped vibrations
(b)   Forced vibrations
(c)   Periodic vibrations

The first two types of vibrations are simple harmonic in nature and the third is non-harmonic in nature.

14.4.1 Equations of vibrations of a simple harmonic type

Equation of linear SHM

Figure 14.1 below shows the movement of a particle executing motion in a circle to illustrate that the point of projection of particle position on the diameter of the circle (or the loot of the perpendicular) is SHM.

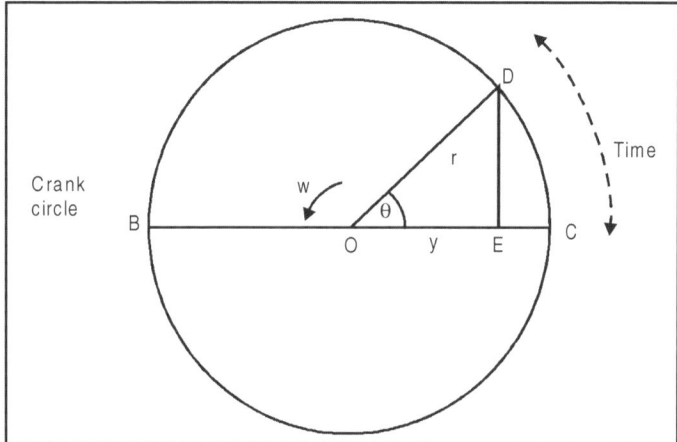

14.1 Representation of a SHM through a circular motion.

In the period of oscillation, a particle travels exactly once around the circle in a given time. Since the angle swept out is a full circle or 2π radians, and the angular velocity is ω radians/s, it follows that the period (T) is given by:

$$T = \frac{2\pi}{\omega}$$

The amplitude of the motion is equal to r, the radius of the circle, and we must now determine how the distance from O to the foot of the perpendicular varies with time. Suppose that, when the particle is at point D, the distance of E (the foot of the perpendicular from D on to BC) from the centre O is y. By geometry:

$$y = r \sin\theta$$

But, if the angular velocity is ω, then the angle θ swept through in a time t, is given by:

$$\theta = \omega t$$

Thus, combining these two formulae, we have an equation for y:

$$y = r \sin \omega t$$

The oscillating motion is thus a sinusoidal one with a characteristic shape. The period of oscillation is the time required for one cycle, the motion from a point on one particular part of one wave to the corresponding point on the next wave, to take place. The number of such cycles occurring in one second is the frequency of the oscillation in Hertz.

The acceleration, f, at E towards O is given by:

$$F = \omega^2 y$$

The correct mathematical expression for the acceleration of a particle undergoing SHM is, then:

$$f = -\omega^2 y$$

The velocity of the particle at D is given as $r\omega$ in a tangential direction. By, geometry, then, the velocity v of E along the diameter BOC can be obtained by vector resolution and is given by:

$$v = r\omega \cos \theta$$

By the well known trigonometrical relation:

$$\sin^2 \theta + \cos^2 \theta = 1$$

We can rewrite the equation of velocity as:

$$v = r\omega\sqrt{\left(1 - \sin^2\theta\right)}$$

However, we have seen that $y = r \sin \grave{e}$, and thus:

$$v = r\omega\sqrt{\left(1 - y^2/r^2\right)}$$

$$v = r\omega\sqrt{\frac{r^2 - y^2}{r^2}}$$

Thus, $$v = \omega\sqrt{\left(r^2 - y^2\right)}$$

The maximum velocity, v_{max} occurs as the particle executing SHM passes through the rest point, and is thus present when y is zero.

Hence, $$v_{max} = \omega r$$

One further useful piece of information is the fact that the maximum and average speeds are related in any SHM and for calculating the amplitude, period, frequency and maximum velocity of its motion. One further important parameter, encountered quite frequently in textiles, is the energy present at a given displacement. In an oscillating spring, this property is particularly important, and this example is convenient one to use in our calculation.

Suppose that a mass m is suspended by a spring, as shown in Fig. 14.2, and reaches an equilibrium position at A where the force of gravity on m is just balanced by the upward force applied by the stretched spring. The stiffness of the spring, expressing the tension that is produced in the spring by extension of unit distance, is σ N/m and the extension at the equilibrium position is d meters.

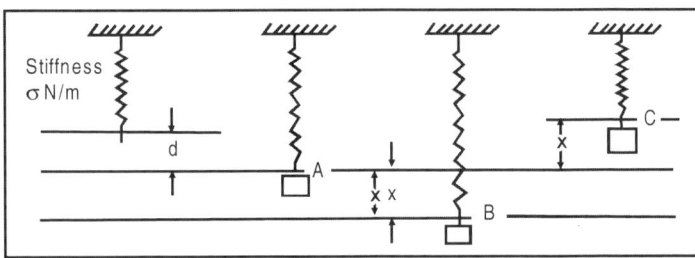

14.2 Mass suspended by spring.

At this equilibrium position, the tension in the spring must just balance the gravitational force acting on the mass, so that:

$$\sigma d = mg$$

Suppose now that the mass is displaced a further distance x meters to a new position B and then released, so that SHM about A occurs between the extreme positions B and C. At B, there is an increase of σx in the tension, while at C there is a corresponding decrease of σx, since AB and AC are equal if damping is ignored. At either extreme position, therefore, there is a force of σx, acting towards the rest position, available for acceleration. Thus the accelerating force acting on the mass is directly proportional to the displacement from the equilibrium position and, in consequence, the same must be true of the acceleration, a fact which confirms that the movement must indeed be SHM.

As the mass rises and falls, a continuous exchange of energy is taking place, though the total energy present remains constant at all times since no external work is done by or on the system. If we take B, the lowest position of the mass, as our datum reference level, all the energy present may then be regarded as strain energy in the spring. In the top position, the strain energy in the spring has been reduced, but has been replaced in the system by a corresponding amount of potential energy as a consequence of the elevation of the mass above the datum reference level. As the mass passes the equilibrium position, it has both potential and kinetic energy, while the spring has an amount of strain energy intermediate between the energies it possesses at the upper and lower positions.

At B the strain energy in the spring is equal to the work done in stretching it by a distance of $(d + x)$ meters. This has already been defined as the total energy, E, of the system, which is thus given by:

$$E = \frac{1}{2}\, \sigma\, (d + x)^2$$
$$= \frac{1}{2}\, \sigma\, d^2 + \frac{1}{2}\, \sigma x^2 + \sigma dx$$

At C, when the mass has risen to its highest position, its potential energy Pc is given by:

$$Pc = mg \times 2x$$
$$= \sigma d \times 2x$$

The strain energy in the spring, Ec, is given by:

$$Ec = \frac{1}{2}\, (d - x)^2$$
$$= \frac{1}{2}\, \sigma d^2 + \frac{1}{2}\, a x^2 - \sigma dx$$

Thus, total energy E in the system, given by $E\,P + E$, is:

$$E = \sigma d \times 2x + \tfrac{1}{2}\sigma d^2 + \tfrac{1}{2}\sigma x^2 - \sigma dx$$
$$= \tfrac{1}{2}\sigma d^2 + \tfrac{1}{2}\sigma x^2 + \sigma dx$$

the same value as in the bottom position.

At the equilibrium position, A, the mass is in motion. The potential energy, P_A is given by:

$$P_A = mgx$$
$$= \sigma dx \text{ as shown earlier in this section.}$$

The strain energy, E_A, is given by:

$$E_A = \tfrac{1}{2}\sigma d^2$$

But the total energy, F, must be the sum of potential, strain and kinetic energies, and must be equal to its former value.

Thus kinetic energy, K_A, must be:

$$K_A = \tfrac{1}{2}\sigma x^2$$

If the velocity of the mass as it passes through A is $v m/s$, an alternative Expression for kinetic energy is:

$$KA = \tfrac{1}{2}mv^2$$

Thus,

$$\tfrac{1}{2}mv^2 = \tfrac{1}{2}\sigma x^2$$

Or:

$$v^2 = \sigma x^2/m$$

This occurs when the velocity has its maximum value, v_{max}, since the equilibrium point is at the centre of the SHM. Thus

$$v_{max} = x''(\sigma/m)$$

We have already shown, however, that:

$$v_{max} = \sigma r$$

Where ω is the angular velocity and r is the radius of the circle from which the SHM is projected. In this case, r is equal to x, the extreme displacement from equilibrium, so that the equivalence:

$$x\sqrt{(\sigma/m)} = \omega r$$

becomes

$$x\sqrt{(\sigma/m)} = \omega x$$

or

$$\omega = \sqrt{(\sigma/m)}$$

The period T is the time taken for one revolution of the generating circle, so that

$$T = 2\pi/\omega$$

Thus:

$$T = 2\pi \sqrt{(m/\sigma)}$$

It is clear that the period of oscillation depends only on the stiffness and the mass, and is completely independent of the amplitude of the oscillation so long as the motion remains simple harmonic in nature.

Equation of angular SHM

Consider the case of a simple pendulum, as shown in Fig. 14.3, consisting of a small mass *m* attached to the end of a light rod of length *l*. If the upper end of the rod is pivoted at the point *P* and the mass is displaced to one side and then released, the system will begin to oscillate in SHM. Suppose, that at a given instant, the rod is inclined at an angle *θ* to the vertical position of equilibrium, as in Fig. 14.3. The forces acting on the mass are the tension, *T*, in the rod and the gravitational attraction vertically downwards. If we resolve forces along the rod, we find:

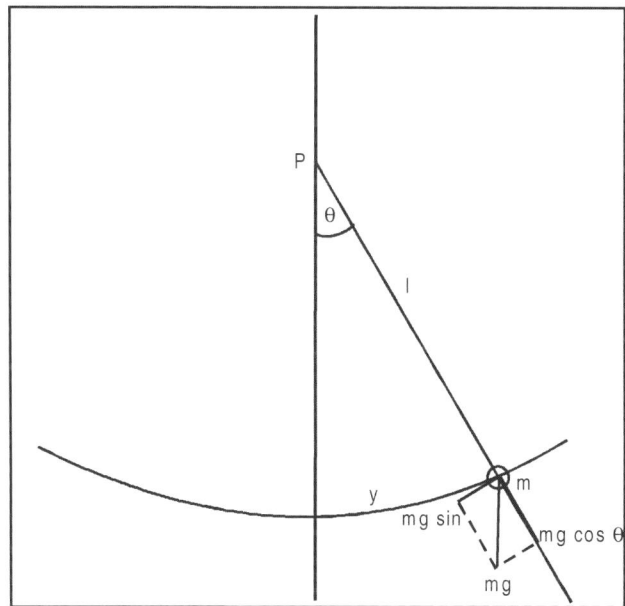

14.3 Particle executing angular SHM.

$$T = g \cos \theta$$

At right angles to the rod, there is only the weight component, *mg* sin *θ*, in this direction so that the system cannot be in equilibrium. This force must therefore produce an acceleration, f, towards the vertical position and, in the tangential direction, we found

$$mg \sin \theta = mf$$

the negative sign being used to indicate that the force is directed towards O whereas displacement is measured positively from O and is therefore in the opposite direction.

In order to demonstrate that SHM is actually occurring, it is now necessary to make the assumption that the maximum displacement is small, and hence that the angle θ is small at all times in the oscillation. Under these circumstances, $\sin \theta$ becomes equal to θ, in radians, and is also equal to y/l, where y is the displacement of the mass from its rest position. With this assumption:

$$mf = -mg \; \theta = -mgy/l$$
i.e.
$$f = -(g/l) \; y$$
$$1 = -(g/l) \; y$$

Since g and l are both constant, it follows that the acceleration f is proportional to the distance y from the rest position, and the oscillation is thus SHM under the specified conditions. The expression can be compared with the corresponding one in linear SHM, where

$$f = -\omega^2 y$$

Combination of the two expressions gives

$$\omega = g/l$$

and, since the period of oscillation, T, is equal to $2\pi/\omega$, we found

$$T = 2\pi \sqrt{(l/g)}$$

The resemblance between linear and angular SHM can be further seen by developing equations for the angular motion, as was done for the linear one. The formula for velocity, for example

$$v = \omega \sqrt{\left(r^2 - y^2\right)}$$

may be rewritten, since $\omega = \sqrt{(g/l)}$, as: $v = \pm\sqrt{\{(g/l)(a^2 - l^2\sin^2 \theta)\}}$
where a is the amplitude, l the length of the pendulum, and θ the instantaneous angular displacement. We know that velocity is zero when the displacement is maximum that is when

$$l \sin \theta = a$$

i.e., when

$$\theta = \sin^{-1} (\sqrt{(g/l)})$$

Similarly, We know that maximum velocity occurs at the mid-point of oscillation, that is, when: $\sin \theta = 0$

i.e. $$v_{max} = \pm \, a\sqrt{(g/l)}$$

the $\pm$ sign indicating that maximum velocity is attained during both directions of travel. The kinetic energy, K, equal to $\frac{1}{2} mv^2$, is given by:

$$K = \frac{mg}{2l} (a^2 - l^2\sin^2 \theta)$$

The total energy of the oscillating system must be experienced when the pendulum bob is at its lowest position, that is, when $\theta = 0$ and v has its maximum value.

Then total energy E is given by

$$E = \frac{mga^2}{2l}$$

and, since total energy E is the sum of the kinetic and potential energies,

the potential energy P is given by $P = \dfrac{mga^2}{2l} - \dfrac{mg}{2l}(a^2 \, l^2\sin^2\theta) = mgl \sin^2\theta$

Other equivalences between linear and angular SHM may be derived in similar manner.

14.4.2 Equations of periodic vibrations

The equation for SHM can be written as

$$y = r \sin \omega t$$

where y is the distance from the rest point at time t, r is the maximum displacement, and ω is the angular velocity of the circular motion theoretically generating the SHM. It can be shown mathematically, by a proof known as Fourier's theorem, that any periodic vibration that repeats regularly can be represented by a series of simple harmonic terms with frequencies that are integral multiples of that of the original function. This means that, if a vibration is repeated at a frequency of n times per second, we can write it down as the resultant of a set of SHM vibrations of frequencies 0, n, $2n$, $3n$, $4n$,... where the angular velocity ω is equal to

2πn. If we take the zero frequency (i.e. constant) term as a reference, we can assign a phase lag ε_1, ε_2, ε_3, ε_4 ... To each subsequent term so that, if the beginning of the zero-frequency signal is taken as zero on the time scale, each of the following terms begins at a time ε_1, ε_2, ε_3, ε_4...etc. later. Thus, the equation of the first harmonic term is

$$y = r_1 \sin (2\pi nt + \varepsilon_1)$$

and, in subsequent terms, n increases by an integer and the phase angle changes. The complete general equation for any non-harmonic periodic vibration can then he written as:

$$y = r_0 + r_1 \sin (2\pi nt + \varepsilon_1) + r_2 \sin (2\pi nt + \varepsilon_2) +$$
$$r_3 \sin (2\pi nt + \varepsilon_3) + r_4 \sin (2\pi nt + \varepsilon_4) +......$$

A standard mathematical transformation enables us to express each of the phase angles ε as a function of the frequency n, so that the expression may be rewritten as:

$$y = r_0 + r_1 \sin (2\pi nt) + b_1 \cos (2\pi nt) + r_2 \sin (2\pi nt) +$$
$$b_2 \cos (2\pi nt) + r_3 \sin (2\pi nt) + b_3 \cos (2\pi nt) +$$
$$r_4 \sin (2\pi nt) + b_4 \cos (2\pi nt) + ...$$

The above equation is applicable to all periodic vibrations.

14.5   Impulse and impulsive force

Impulse = force × time

Now consider a body of mass m. Let a force F change its velocity from an initial velocity v_1 to a final velocity v_2.

Force is equal to the rate of change of linear momentum, therefore

$$F = \frac{m (v_2 - v_1)}{t} \text{ or } F \times t = m (v_2 - v_1)$$

i.e., Impulse = change of linear momentum, where F is force and t is time

14.6   Collision of bodies

14.6.1 Collision of inelastic bodies

Momentum before impact = Momentum after impact
$$m_1 \bullet u_1 + m_2 \bullet u_2 = (m_1 + m_2)v$$

$$v = \frac{m_1 \cdot u_1 + m_2 \cdot u_2}{m_1 + m_2}$$

m_1 – mass of first body

m_2 – mass of second body

u_1 and u_2 – velocities of first and second bodies before impact

v – common velocity of first and second bodies after impact

14.6.2 Collision of elastic bodies

$$\text{Coefficient of restitution} = \frac{\text{Relative velocity after impact}}{\text{Relative velocity before impact}}$$

14.7   Friction

It is a well known fact that the surfaces of the bodies are never perfectly smooth. In other words even a body having a smooth surface is found to have roughness and irregularities, which may not be detected by ordinary touch. If a block of one substance be placed over the level surface of the same or of different material, a certain degree of interlocking of the minutely projecting particles takes place. This does not involve any force, so long as the block does not move or tends to move. But whenever one block moves or tends to move tangentially with respect to the surface, on which it rests, the interlocking property of the projecting particles opposes the motion. This opposing force, which acts in the opposite direction of the movement of the upper block, is called the *force of friction* or simply *friction*. It thus follows that at every joint in a machine, force of friction arises due to the relative motion between two parts and hence some energy is wasted in overcoming the friction. Though the friction is considered undesirable, yet it plays as important role both in nature and in engineering e.g. walking on a road, motion of locomotive on rails, transmission of power by belts, gears etc.

14.7.1 Types of friction

In general, the friction is of the following two types:

(1) Static friction

It is the friction, experienced by a body, when at rest.

(2) Dynamic friction

It is the friction, experienced by a body, when in motion. The dynamic friction is also called *kinetic friction* and is less than the static friction. It is of the following three types:

(a) Sliding friction

It is the friction, experienced by a body, when it slides over another body.

(b) Rolling friction

It is the friction, experienced between the surfaces which has balls or rollers interposed between them.

(c) Pivot friction

It is the friction, experienced by a body, due to the motion of rotation as in case of foot step bearings.

14.7.2 Limiting friction or limiting value of friction

It is the maximum value of frictional force, which comes into play, when a body just begins to slide over the surface of another body. It may be noted that when the applied force is less than the limiting friction, the body remains at rest, and the friction into play is called static friction which may have any value between zero and limiting friction.

14.7.3 Laws of static friction

(i)   The force of friction always acts in a direction, opposite to that in which the body tends to move.
(ii) The magnitude of the force of friction is exactly equal to the force, which tends the body to move.
(iii) The magnitude of the limiting friction bears a constant ration to the normal reaction between the two surfaces
(iv) The force of friction is independent of the area of contact, between the two surfaces.
(v)   The force of friction depends upon the roughness of the surfaces.

14.7.4 Laws of kinetic or dynamic friction

(i)   The force of friction always acts in a direction, opposite to that in which the body is moving.

(ii) The magnitude of the kinetic friction bears a constant ratio to the normal reaction between the two surfaces. But this ratio is slightly less than that in the case of limiting friction.

(iii) For moderate speeds, the force of friction remains constant. But it decreases slightly with the increase of speed.

14.7.5 Co-efficient of friction

It is defined as the ratio of the limiting friction to the normal reaction between the two bodies.

$$F = \mu \cdot W = \mu \cdot R_N$$

Where F = frictional force

μ = coefficient of friction

R_N = normal reaction

W = applied load

$$\mu = F/R_N$$

Minimum force required to slide a body on a rough horizontal plane

$$P = \frac{W \sin \varphi}{\cos(\theta - \varphi)}$$

Where P = minimum force

φ = limiting angle of friction

θ = angle at which load is applied to the horizontal

14.8   Belt and gear drives

Belts or ropes are used to transmit power from one shaft to another by means of pulleys which rotate at the same speed or at different speeds. The amount of power transmitted depends upon the following factors

(i) The velocity of the belt.

(ii) The tension under which the belt is placed on the pulleys.

(iii) The arc of contact between the belt and the smaller pulley.

(iv) The conditions under which the belt is used.

14.8.1 Types of belts

The following are the types of belts in common use:

(i) Flat belt

This type of belt is used where a moderate amount of power is to be transmitted, from one pulley to another when the two pulleys are not more than 8 meters apart.

(ii) V-belt

This type of belt is used where a moderate amount of power is to be transmitted, from one pulley to another, when the two pulleys are very near to each other

(iii) Circular belt or rope

This type of belt is used where a great amount of power is to be transmitted, from one pulley to another, when the two pulleys are more than 8 meters apart.

When a large amount of power is to be transmitted, then a single belt may not be sufficient. In such a case, wide pulleys (for V-belts or circular belts) with a number of grooves are used. Then a belt in each groove is provided to transmit the required amount of power from one pulley to another.

14.8.2 Types of belts

The power from one pulley to another may be transmitted by any of the following types of belt drives.

(i) *Open belt drive*. The open belt drive, as shown in Fig. 14.4, is used with shafts arranged parallel and rotating in the same direction. In this

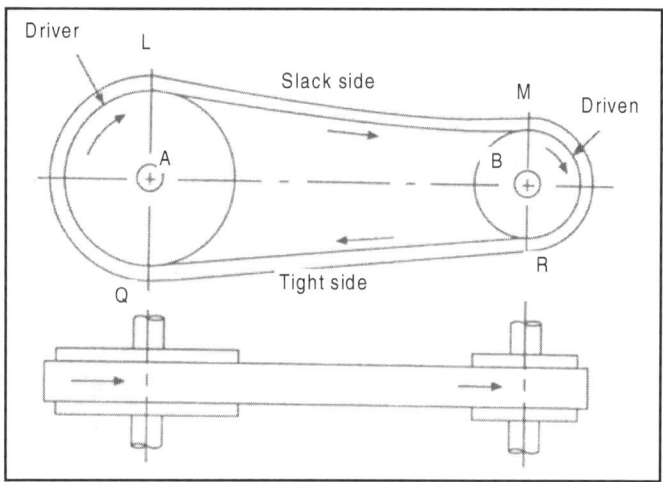

14.4 Open belt drive.

case, the driver A pulls the belt from one side (i.e. lower side *RQ*) and delivers it to the other side, i.e. upper side *LM*). Thus the tension in the lower side belt will be more than that in the upper side belt. The lower side belt (because of more tension) is known as tight side whereas the upper side belt (because of less tension) is known as slack side as shown in Fig. 14.4.

(ii) *Crossed or twist belt drive*. The crossed or twist belt drive, as shown in Fig. 14.5, is used with shafts arranged parallel and rotating in the opposite directions. In this case, the driver pulls the belt from one side (i.e. RQ) and delivers it to the other side (i.e. LM). Thus the tension in the belt RQ will be more than that in the belt *LM*. The belt *RQ* (because of more tension) is known as tight side, whereas the belt *LM* (because of less tension) is known as slack side, as shown in Fig. 14.5.

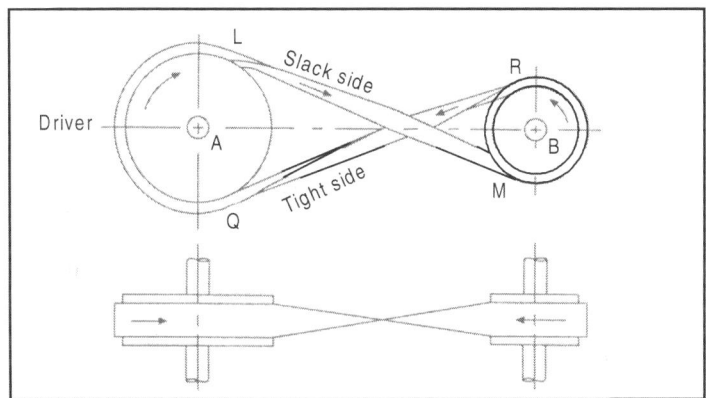

14.5 Crossed belt drive.

(iii) *Quarter turn belt drive*. The quarter turn belt drive also known as right angle belt drive, as shown in Fig. 14.6, is used with shafts arranged at right angles and rotating in one definite direction. In order to prevent the belt from leaving the pulley, the width of the face of the pulley should be greater or equal to 14.4 b, where b is the width of belt.

In case the pulleys cannot be arranged, as shown in Fig. 14.6, or when the reversible motion is desired, then a quarter turn belt drive with guide pulley, as shown in Fig. 14.6, may be used.

(iv) *Belt drive with idler pulleys*. A belt drive with an idler pulley, as shown in Fig. 14.7, is used with shafts arranged parallel and when an open belt drive cannot be used due to small angle of contact on the smaller pulley. This type of drive is provided to obtain high velocity ratio and when the required belt tension cannot be obtained by other means.

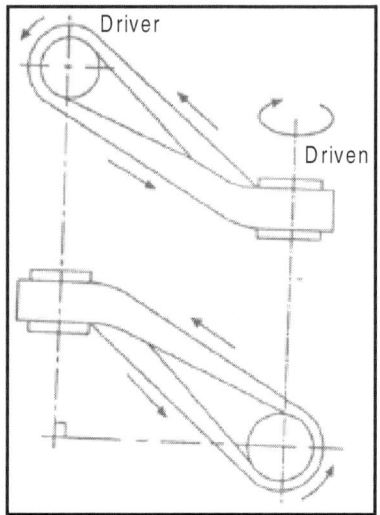

14.6 Quarter turn belt drive.

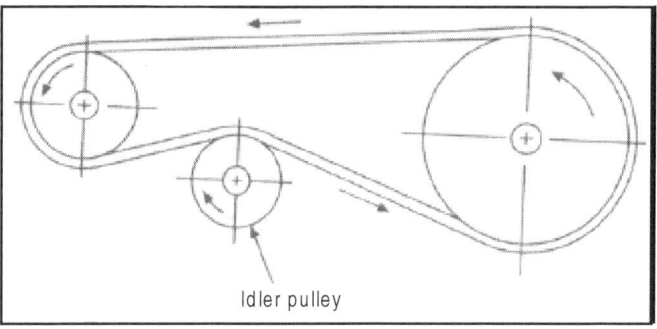

14.7 Belt drive with an idler pulley

When it is desired to transmit motion from one shaft to several shafts, all arranged in parallel, a belt drive with many idler pulleys, as shown in Fig. 14.8, may be employed.

(v) *Compound belt drive.* A compound belt drive, as shown in Fig. 14.9, is used when power is transmitted from one shaft to another through a number of pulleys.

(vi) *Stepped or cone pulley drive.* A stepped or cone pulley drive, as shown in Fig. 14.10 is used for changing the speed of the driven shaft while the main or driving shaft runs at constant speed. This is accomplished by shifting the belt from one part of the steps to the other.

(vii) Fast and loose pulley drive. Fast and loose pulleys drive, as shown if Fig. 14.11 is used when the driven or machine shaft is to be started or stopped when ever desired without interfering with the driving shaft. A

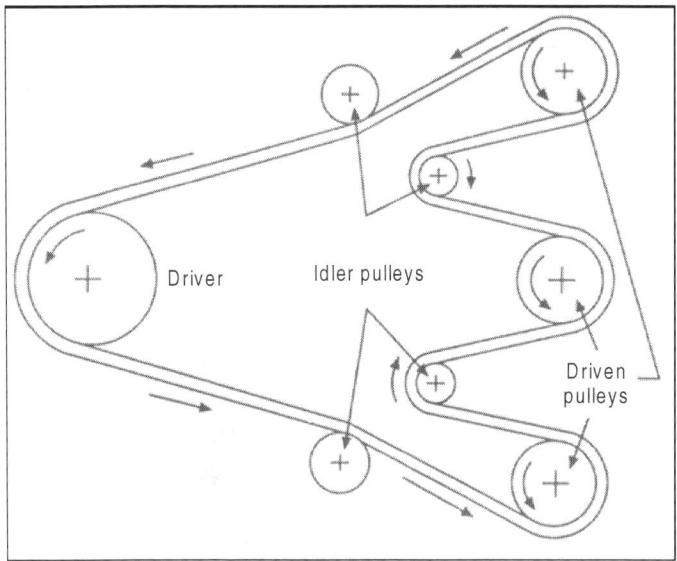

14.8 Belt drive with many idler pulleys.

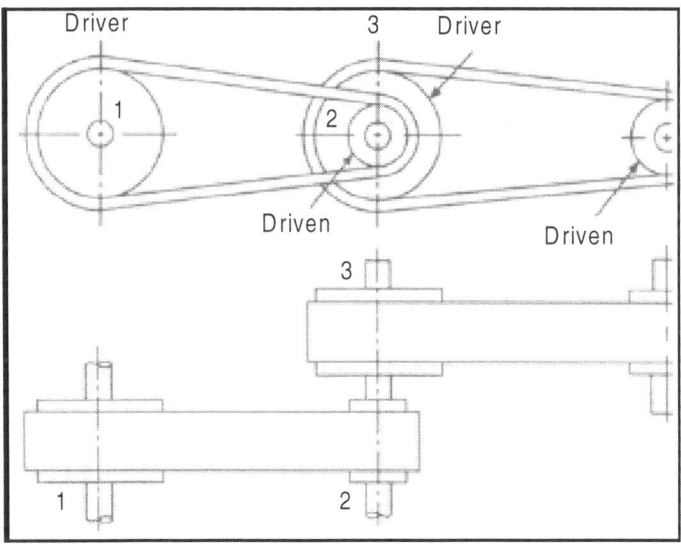

14.9 Compound belt drive.

pulley which is keyed to the machine shaft is called fast pulley and runs at the same speed as that of machine shaft. A loose pulley runs freely over the machine shaft and is incapable of transmitting any power. When the driven shaft is required to be stopped, the belt is pushed on to the loose pulley by means of sliding bar having belt forks.

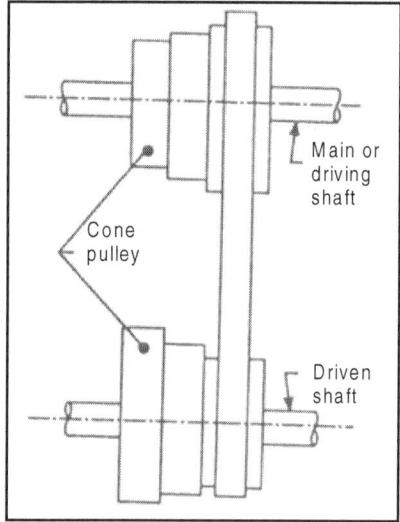

14.10 Stepped or cone pulley drive.

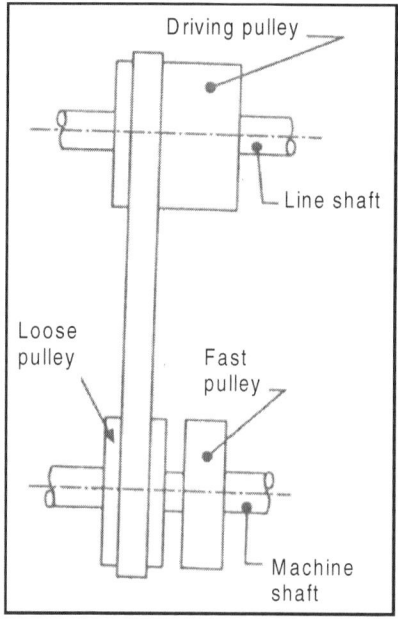

14.11 Fast and loose pulley drive.

Velocity ratio of a simple flat belt drive

$$\frac{N_2}{N_1} = \frac{d_2}{d_1}$$

or

$$\frac{N_2}{N_1} = \frac{d_2 + t}{d_1 + t}$$

Where d_1 = diameter of the driver pulley
d_2 = diameter of the follower pulley
N_1 = speed of driver in rpm
N_2 = speed of driven in rpm
t = thickness of belt

Velocity ratio of a compound flat belt drive

$$\frac{N_4}{N_1} = \frac{d_1 \times d_3}{d_2 \times d_4}$$

Where d_1 = diameter of pulley 1,
N_1 = speed of the pulley 1 in rpm,
d_2, d_3, d_4 and N_2, N_3, N_4 – corresponding values for pulleys 2, 3 and 4.

Or $\dfrac{\text{speed of last driven}}{\text{speed of first driver}} = \dfrac{\text{product of diameters of drivers}}{\text{product of diameters of driven}}$

Slip of belt is given by $\dfrac{N_2}{N_1} = \dfrac{d_1(1 - s_1 + s_2)}{d_2} = \dfrac{d_1(1 - s)}{d_2 \times 100}$

Where $\sigma = \sigma_1 + \sigma_2$, i.e. total percentage of slip
If thickness of belt (t) is considered, then

$$\frac{N_2}{N_1} = \frac{d_1 + t(1 - s)}{d_2 + t(100)}$$

Creep of belt

$$\frac{N_2}{N_1} = \frac{d_1}{d_2} \times \frac{E + \sqrt{\sigma_2}}{E + \sqrt{\sigma_1}}$$

where σ_1 and σ_2 are stresses in the belt on the tight and slack side respectively, and E is Young's modulus for the material of the belt.

Power transmitted by a belt

$$P = (T_1 - T_2) \, v \text{ Watts}$$

Where P = power transmitted
T_1 = Tension on tight side of the belt,
T_2 = Tension on the slack side of the belt,
v = velocity of the belt in m/s

Ratio of driving tensions for a flat belt drive

$$2.3 \log [T_1/T_2] = \mu.\theta$$

where θ angle of contact in radians between belt and pulley

Centrifugal tension of a flat belt

$$Tc = m.v^2$$

Where Tc = centrifugal tension of the belt
m = mass of the belt per unit length in kg,
v = linear velocity of the belt in m/s

Ratio of driving tensions for V-belt

$$2.3 \log [\, T_1/T_2] = \mu . R \, cosec \, \beta$$

Where μ is coefficient of friction between the belt and siles of the
groove,
R is total reaction in the plane of the groove, and
β is angle of the groove of the pulley

14.9   Gears

In the case of belt or rope drives, the slippage is a common phenomenon, in the transmission of motion or power between two shafts. The effect of slipping is to reduce the velocity ratio of the system. In precision machines, in which a definite velocity ration is of importance, the only positive drive is by means of gears or toothed wheels. A gear drive is also provided, when the distance between the driver and the follower is very small.

14.9.1 Merits and demerits

The following are the merits and demerits of the gear drive as compared to belt, rope and chain drives:

Merits

(i)   It transmits exact velocity ratio.
(ii) It may be used to transmit large power.
(iii) It has high efficiency.
(iv) It has reliable service.
(v)   It has compact layout.

Demerits

(i)   The manufacture of gears requires special tools and equipment.
(ii) The error in cutting teeth may cause vibrations and noise during operation.

14.9.2 Classification of gears

The gears or toothed wheels may be classified as follows:

(i) According to the position of axes of the shafts
    (a)   Parallel (spur gear),
    (b)   Intersecting (helical gear), and
    (c)   Non-intersecting and non-parallel (Bevel and spiral gear).

(ii) According to the peripheral velocity of the gears. The gears, according to the peripheral velocity of the gears may be classified as
    (a)   Low velocity,
    (b)   Medium velocity, and
    (c)   High velocity

(iii) According to the type of gearing. The gears, according to the type of gearing may be classified as:
    (a)   External gearing,
    (b)   Internal gearing, and
    (c)   Rack and pinion

(iv) According to the position of the teeth on the gear surface. The teeth on the gear surface may be
    (a)   straight,
    (b)   inclined, and
    (c)   curved

$$\text{Circular pitch, } p_c = \pi D/T$$

Where D = diameter of the pitch circle
          T = number of teeth on the wheel

$$\text{Diametrical pitch} = T/D = ir/p$$

Where T = number of teeth, and
          D = pitch circle diameter
    Module = D/T

14.9.3 Gear trains

Two or more gears can be made to mesh with each other to transmit power from one shaft to another. Such a combination is called gear train or train of toothed wheels. The nature of the train used depends upon the velocity ration required and the relative position of the axes of the shafts. A gear train may consist of spur, bevel or spiral gears.

14.9.4 Classification of gear trains

Gear trains are classified as follows:

(i)   Simple gear train,
(ii) Compound gear train,
(iii) Reverted gear train, and
(iv) Epicyclical gear train

In the first three types of gear trains, the axes of the shafts over which gears are mounted are fixed relative to each other. But in the case of epicyclical gear trains, the axes of the shafts on which the gears are mounted may move relative to a fixed axis.

14.9.5 Simple gear train

When there is only one gear on each shaft, as shown in Fig. 14.12, it is known as simple gear train. The gears are represented by their pitch circles.

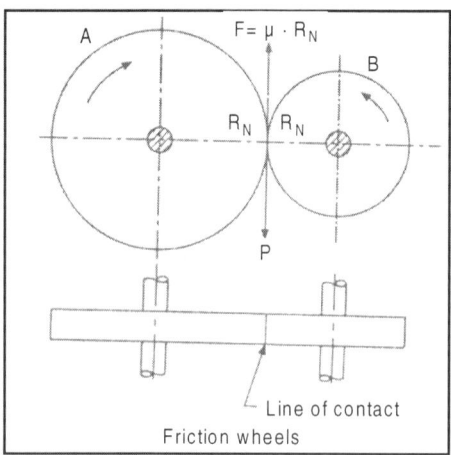

14.12 Simple gear train.

When the distance between the two shafts is small, the two gears 1 and 2 are made to mesh with each other to transmit motion from one shaft to the other, as shown in Fig. 14.12. Since the gear 1 drives the gear 2, therefore gear 1 is called the driver and the gear 2 is called the driven or follower. It may be noted that the motion of the driven gear is opposite to the motion of driving gear.

Since the speed ratio or velocity ratio of gear train is the ratio of the speed of the driver to the speed of the driven or follower and ratio of speeds of any pair of gears in mesh is the inverse of their number of teeth, therefore

$$\text{Speed ratio of a simple gear train} = \frac{N_1}{N_2} = \frac{T_2}{T_1}$$

It may be noted that the ratio of the speed of the driven or follower to the speed of the driver is known as train value of the gear train. Mathematically

$$\text{Train} = \frac{N_2}{N_1} = \frac{T_1}{T_2}$$

Where N_1 = speed of driver gear in rpm
N_2 = speed of driven gear in rpm
T_1 = number of teeth on driver gear
T_2 = number of teeth on driven gear

Thus it can be seen that the train value is the reciprocal of the speed ratio. Sometimes, the distance between the two gears is large. The motion from one gear to another, in such a case, may be transmitted by either of the following two methods:

(a) By providing the large sized gear, or
(b) By providing one or more intermediate gears.

The first method is very inconvenient and uneconomical; whereas the second method is very convenient and economical.

It may be noted that when the number of intermediate gears are odd, the motion of both the gears (i.e. driver and driven or follower) is as shown in Fig. 14.12.

But if the numbers of intermediate gears are even, the motion of the driven or follower will be in the opposite direction of the driver as shown in Fig. 14.12.

Now consider a simple train of gears with one intermediate gear as shown in Fig. 14.12.

Let N_1 = speed of driver in rpm
N_2 = speed of driven in rpm
N_3 = speed of driven or follower in rpm
T_1 = number of teeth on driver
T_2 = number of teeth on intermediate gear, and
T_3 = number of teeth on driven or follower.

Since the driving gear 1 is in mesh with the intermediate gear 2, therefore speed ratio for these two gears is

$$\frac{N_1}{N_2} = \frac{T_2}{T_1} \qquad \cdots \text{(i)}$$

Similarly, as the intermediate gear 2 is in mesh with the driven gear 3, therefore speed ratio for these two gears is

$$\frac{N_2}{N_3} = \frac{T_3}{T_2} \qquad \cdots \text{(ii)}$$

The speed ratio of the gear train is obtained by multiplying the equations (i) and (ii).

Therefore, $$\frac{N_1}{N_2} \times \frac{N_2}{N_3} = \frac{T_2}{T_1} \times \frac{T_3}{T_2}$$

or $$\frac{N_1}{N_3} = \frac{T_3}{T_1}$$

i.e. $\dfrac{\text{speed of driver = number of teeth on driven}}{\text{speed of driven = number of teeth on driver}}$

and Train value = $\dfrac{\text{speed of driven}}{\text{speed of driver}} = \dfrac{\text{number of teeth on driver}}{\text{number of teeth on driven}}$

Similarly, it can be proved that the above equation holds good even if there are any number of intermediate gears. From above, we see that the speed ratio and the train value, in a simple train of gears, are independent of the size and number of intermediate gears. These intermediate gears are called idle gears, as they do not affect the speed ratio or train value of the system. The idle gears are used for the following two purposes:

(i)   to connect gears where a large centre distance is required, and
(ii) to obtain the desired direction of motion of the driven gear (i.e. clockwise or anticlockwise)

14.9.6 Compound gear train

In this type of gear train, there is more than one gear on a shaft. It has been seen earlier that the idle gears in a simple train of gears do not affect the speed ratio of the system. But these gears are useful in bridging over the space between the driver and the driven. But whenever the distance between the driver and the driven has to be bridged over by intermediate gears and at the same time a large or small speed ratio is required, then the advantage of intermediate gears is intensified by providing compound gears on intermediate shafts. In this case, each intermediate shaft has two gears rigidly fixed to it so that they may have the same speed. One of these two gears meshes with the driver and the other with the driven attached to the next shaft as shown in Fig. 14.13.

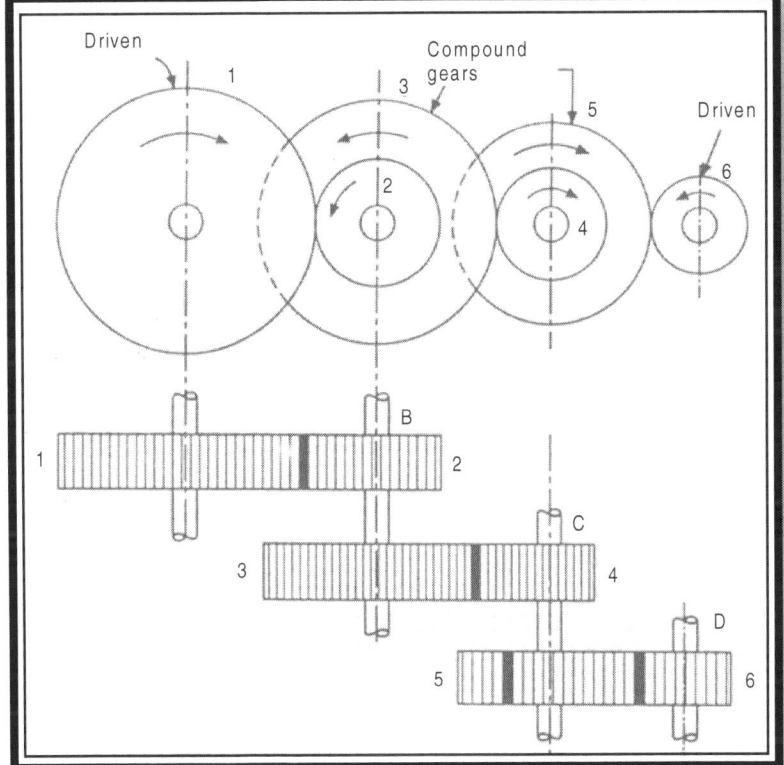

14.13 Compound gear train.

In a compound train of gears, as shown in Fig. 14.13, the gear 1 is the driving gear mounted on shaft A, gears 2 and 3 are compound gears which are mounted on shaft B. The gears 4 and 5 are also compound gears which are mounted on shaft C and the gear 6 s the driven gear mounted on shaft *D*.

Let N_1 = speed of driving gear 1,

T_1 = number of teeth on driving gear 1,

N_2, N_3 ... , N_6 = speed of respective gears in rpm, and

T_2, T_3 ..., T_6 = number of teeth on respective gears.

Since gear 1 is in mesh with gear 2, therefore its speed ratio is

$$\frac{N_1}{N_2} = \frac{T_2}{T_1} \qquad \dots \text{(i)}$$

Similarly, for gears 3 and 4, speed ratio is

$$\frac{N_3}{N_4} = \frac{T_4}{T_3} \qquad \dots \text{(ii)}$$

And for gears 5 and 6, speed ratio is

$$\frac{N_5}{N_6} = \frac{T_6}{T_5} \qquad \dots \text{(iii)}$$

The speed ratio of compound gear train is obtained by multiplying the equations (i), (ii) and, (iii)

Therefore $$\frac{N_1}{N_2} \times \frac{N_3}{N_4} \times \frac{N_5}{N_6} = \frac{T_2}{T_1} \times \frac{T_4}{T_3} \times \frac{T_6}{T_5}$$

or* $$N_1 = \frac{T_2}{T_1} \times \frac{T_4}{T_3} \times \frac{T_6}{T_5}$$

* Since gears 2 and 3 are mounted on one shaft B, therefore $N_2 = N_3$. Similarly gears 4 and 5 are mounted on shaft C, therefore $N_4 = N_5$.

$$\text{Speed ratio} = \frac{\text{product of the number of teeth on the drivers}}{\text{product of the number of teeth on the driven}}$$

$$= \frac{\text{product of the number of teeth on the driven}}{\text{product of the number of teeth on the drivers}}$$

$$\text{Train value} = \frac{\text{speed of the last driven or follower}}{\text{speed of the first driver}}$$

$$= \frac{\text{product of the number of teeth on the drivers}}{\text{product of the number of teeth on the driven}}$$

The advantage of a compound train over a simple gear train is that a much large speed reduction from the first shaft to the last shaft can be obtained with small gears. If a simple gear train is used to give a large speed reduction, the last gear has to be very large. Usually for a speed reduction in excess of 7 to 1, a simple train is not used and compound train or worm gearing is employed.

14.9.7 Reverted gear train

When the axes of the first gear (i.e. first driver) and the last gear (i.e. last driven or follower) are co-axial, then the gear train is known as reverted gear train as shown in Fig. 14.14.

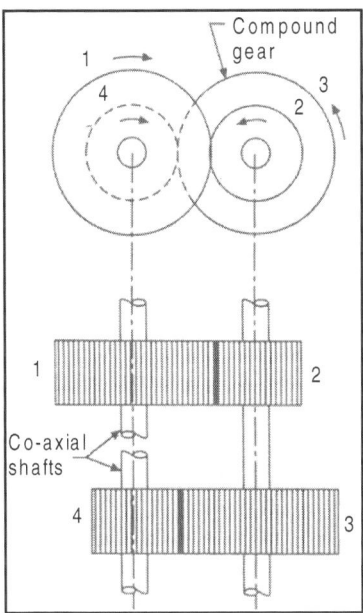

14.14 Reverted gear train.

It can be seen that gear 1 drives the gear 2 in the opposite direction. Since the gear 2 and 3 are mounted on the same shaft, therefore they form a compound gear and the gear 3 will rotate in the same direction as that of gear 2. The gear 3 (which is now the second driver) drives the gear 4 (i.e. the last driven or follower) in the same direction that of gear 1. Thus we see that in a reverted gear train, the motion of the first gear and the last gear is similar.

Let T_1 = number of teeth on gear 1,
        r_1 = pitch circle radius of gear 1, and
N_2, N_3, N_1 = speed of gear 1 in rpm

Similarly,

T_2, T_3, T_4 = number of teeth on respective gears,
r_2, r_3, r_4 = pitch circle radii of respective gears, and
N_2, N_3, N_4 = speed of respective gears in rpm.

Since the distance between the centers of the shafts of gears 1 and 2 as well as gears 3 and 4 is same, therefore

$$r_1 + r_2 = r_3 + r_4 \qquad \ldots \text{(i)}$$

Also, the circular pitch or module of all the gears is assumed to be same; therefore number of teeth on each gear is directly proportional to its circumference or radius

Therefore $\qquad *\ T_1 + T_2 = T_3 + T_4 \qquad \ldots \text{(ii)}$

$*$ Since the circular pitch, $P_c = \dfrac{2\pi r}{T} = \pi m$ or $r = \dfrac{m \cdot T}{2}$ where m is the module.

Therefore $\ r_1 = \dfrac{m \cdot T_1}{2} ; r_2 = \dfrac{m \cdot T_2}{2} ; r_3 = \dfrac{m \cdot T_3}{2} ; r_4 = \dfrac{m \cdot T_4}{2}$

Now from equation (i),

$$\frac{m \cdot T_1}{2} + \frac{m \cdot T_2}{2} + \frac{m \cdot T_3}{2} + \frac{m \cdot T_4}{2}$$

Thus $T_1 + T_2 = T_3 + T_4$

$$\text{The speed ratio} = \frac{\text{product of number of teeth on driven}}{\text{product of number of teeth on drivers}}$$

Or $\qquad \dfrac{N_1}{N_4} = \dfrac{T_2}{T_1} \times \dfrac{T_4}{T_3} \qquad \ldots \text{(ii)}$

From equations (i), (ii) and (iii), we can determine the number of teeth on each gear for the given centre distance, speed ratio and module only when the number of teeth o one gear is chosen arbitrarily.

14.9.8 Epicyclic gear train

In an epicyclic gear train, the axes of the shafts, over which the gears are mounted, move relative to a fixed axis. A simple epicyclic gear train is shown in Fig. 14.15. L where a gear A and the arm C have a common axis at O_1 about which they can rotate. The gear B meshes with gear A and has its axis on the arm O_2, about which the gear B can rotate. If the arm is fixed, the gear train is simple and gear A can drive gear B or vice versa, but if gear A is fixed and the arm is rotated about the axis of gear A (i.e. O_1) then the gear B is forced to rotate upon and around gear A. Such a motion is called epicyclic and the gear trains arranged in such a manner

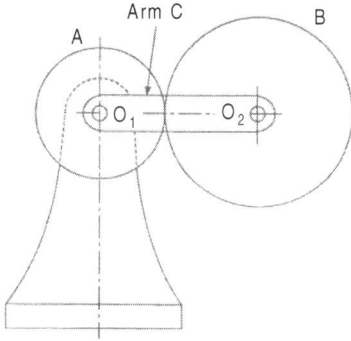

14.15 Epicyclic gear train.

that one or more of their members move upon and around another member are known as epicyclic gear trains (epi means upon and cyclic means around). The epicyclic gear trains may be simple or compound.

The epicyclic gear trains are useful for transmitting high velocity ratios with gears of moderate size in a comparatively lesser space. They are used as differential gears in the speed frames, for driving bobbins and in the Roper positive let off in looms.

14.9.9 Velocity ratio of epicyclic gear train

The following two methods may be used for finding out the velocity ratio of an epicyclic gear train: (a) Tabular method and (b) Algebraic method

(a) Tabular method

Consider an epicyclic gear train as shown in Fig. 14.15

Let T_A = number of teeth on gear A, and
T_B = number of teeth on gear B.

Assume that the arm is fixed. Therefore the axes of both the gears are also fixed relative to each other. When the gear A makes one revolution anticlockwise, the gear B will make T_A/T_B revolutions clockwise. Assuming the anticlockwise rotation as positive arid clockwise as negative, we may say that when gear A makes +1 revolution, then the gear B will make $(-T_A/T_B)$ revolutions. This statement of relative motion is entered in the first row of Table 14.1.

Next, if the gear A makes +X revolutions, then the gear B will make

$$-X \cdot \frac{T_A}{T_B}$$ revolutions.

This statement is entered in the second row of the Table 14.1. In other words, multiply the each motion (entered in the first row) by X.

Table 14.1 Table of motions

Step no.	Conditions of motion	Revolutions of elements		
		Arm C	Gear A	Gear B
1.	Arm fixed – gear A rotates through +1 revolution i.e. 1 rev. anticlockwise	0	+ 1	$-\dfrac{T_A}{T_B}$
2.	Arm fixed – gear A rotates through +X revolutions	0	+X	$-X \cdot \dfrac{T_A}{T_B}$
3.	Add +Y revolutions to all elements	+Y	+Y	+Y
4.	Total motion	+Y	X + Y	$Y - X \cdot \dfrac{T_A}{T_B}$

Next, each element of an epicyclic train is given $+Y$ revolutions and entered in the third row. Finally, the motion of each element of the gear train is added up and entered in the fourth row. A little consideration will show that when the two conditions about the motion of rotation of any two elements are known, then the unknown speed of the third element may he obtained by substituting the given data in the third column of the fourth row.

(b) Algebraic method

In this method, the motion of each element of the epicyclic train relative to the arm is set down in the form of equations. The number of equations depends upon the number of elements in the gear train. But the two conditions are, usually, supplied in any epicyclic train viz., some element is fixed and the other has specified motion. These two conditions are sufficient to solve all the equations; and hence to determine the motion of any element in the epicyclic gear train.

Let the arm be C be fixed in an epicyclic gear train as shown in Fig. 14.15. Therefore speed of the gear A relative to the arm C.

$$= N_A - N_C$$

and, the speed of the gear B relative to the arm C,

$$= N_B - Nc$$

Since the gears A and B are meshing directly, therefore they will revolve in opposite directions.

Therefore $$\dfrac{N_B - Nc}{N_A - N_C} = -\dfrac{T_A}{T_B}$$

Since the arm C is fixed, therefore its speed, $Nc = 0$. If the gear A is fixed, then $N_A = 0$.

Therefore

$$\frac{N_B - Nc}{0 - N_C} = -\frac{T_A}{T_B}$$

or

$$\frac{N_B}{N_C} = 1 + \frac{T_A}{T_B}$$

Note: Of the above two methods, the tabular method is easier and hence mostly used in solving problems on epicyclic gear train.

14.9.10 Bevel wheels

These wheels, shown in Fig. 14.16, are used to transmit motion to shafts perpendicular to each other. Their calculations are similar to those of spur gears. These gears are used in the mill warping machines, sizing machines, etc.

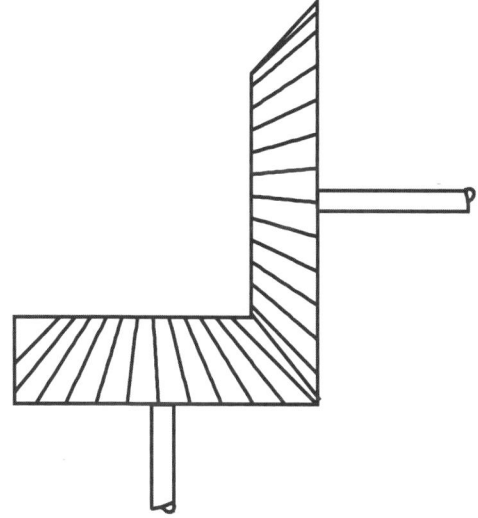

14.16 Bevel wheels.

14.9.11 Worm and worm wheel

These are used to transmit motion to perpendicular shafts. Worm is used only as a driver and in a case where the speed is to be considerably diminished. In calculation a single thread worm is taken as a wheel of one tooth, a double thread worm as a wheel of 2 teeth and so on. The wheel which is driven by a worm is called a worm wheel. In Fig. 14.17, A is worm and B is a worm wheel. The use of these gears may be seen in

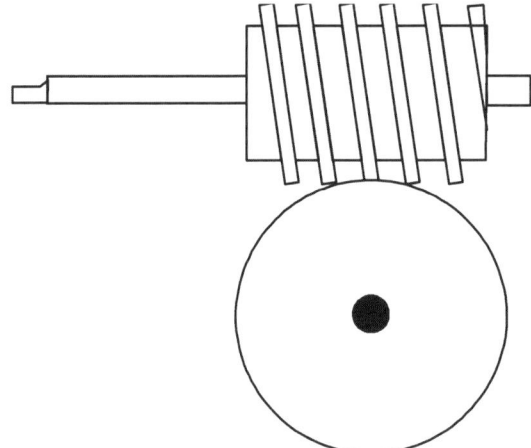

14.17 Worm and worm wheel.

measuring motion of beam warping and slasher sizing machines and also in continuous type of positive take up motion in looms.

14.9.12 Ratchet wheel

Ratchet wheel, as shown in Fig. 14.18, is used to convert an intermittent to and fro motion into an intermittent circular motion as in the take up motions of looms.

14.9.13 Mangle wheel

This wheel, as shown in Fig. 14.19, is used only as a driven wheel. It reverses its motion automatically. The tooth at each end is used only once in a double revolution. Therefore in calculation the wheel is considered as a wheel containing one tooth less than its actual number of teeth. Mangle wheel motion is used in some of the ordinary types of upright spindle winding as also in pirn winding machines.

14.10   Some important quantitative definitions

Ounce (Ozs)

It is the weight equivalent to 28.15 g and is 1/16th of a pound.

Pound (lbsi)

It the weight equivalent to 453.6 g. One kg weight equals 2.2 lbs.

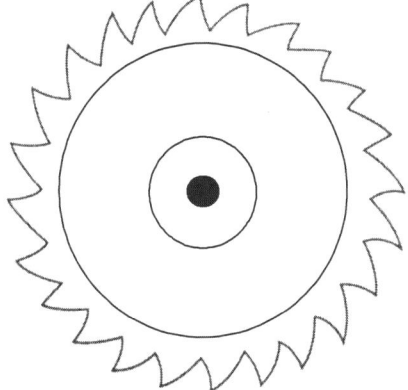

14.18 Ratchet wheel

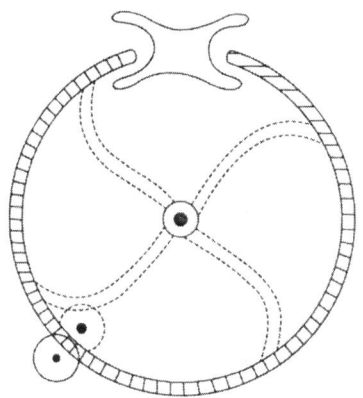

14.19 Mangle wheel.

Grain

It is the weight equivalent to 1/7000th of a pound (1 lb 7000 grains).

Hank

It is the number of times of length of 840 yards contained in one pound of a material.

For example, 840 yards of material contained in a pound is one hank and 8400 yards contained in a pound is 10 hanks.

Lea/skein

It is the length of 120 yards.

Calculations in fibre testing

15.1   Introduction

This chapter deals with formulae used in fibre testing. Some key definitions are given to give a better understanding of the formulae used.

15.2   Basic definitions

The following definitions can be explained by means of the diagram (Fig. 15.1) shown below.

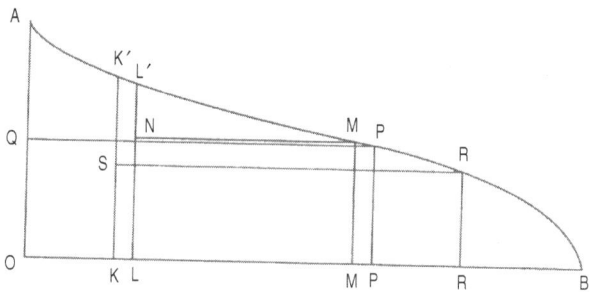

15.1 Sorter diagram.

Average or mean fibre length

It is the sum of the base line readings divided by the base line length and is given in units of 1/8 in., the class interval.

Maximum fibre length

This is read directly from the diagram.

Modal length

The mode of a frequency distribution is the class value of the class with the highest frequency. Since distances along the base line are assumed to be proportional to frequency, the modal length will be given by the class which has the greatest base length.

Effective fibre length

This is the length of the main bulk of the longer fibres. It is obtained by a geometrical construction on the sorter diagram.

Percentage short fibre

This is the percentage of fibres less than half the effective length. As per the figure above, the percentage of short fibre is given by

[RB/OB] × 100 (per cent)

Dispersion

The uniformity or variability of the cotton, can be expressed as the 'inter quartile range'. In figure $L'L$ is the upper quartile and $M'M$ is the lower quartile, where $OM = 3/4$ *or* the difference between $L'L$ and $M'M$, NL' is the inter-quartile range. This measure of the dispersion may now be expressed as a percentage of the effective length:

Dispersion = [$NL/L'L$] × 100 (per cent)

Fibre fineness

It is measure of the linear density of the fibres and is normally expressed in terms of weight per unit length or micronaire.

Maturity ratio

It is the ratio which expresses the actual fibre weight per cm in relation to a standard fibre weight per cm. Standard fibre weight per cm is that which the fibre would have if it were fully matured in the sense of having an $N - D$ of 60. (N is the number of normal fibres and D is the number of dead fibres)

15.3   Various other formulae

Relative humidity, RH = [actual vapour pressure / saturated vapour pressure] × 100

Moisture content, $M = \dfrac{\text{Weight of water}}{\text{Total weight of the material}} \times 100$

$$\text{Moisture regain, } R = \frac{\text{Weight of water}}{\text{Oven dry weight of the material}} \times 100$$

Relation between M and R

$$R = \frac{100W}{D} \qquad\qquad M = \frac{100W}{D + W}$$

$$R = \frac{M}{\frac{1 - M}{100}} \qquad\qquad M = \frac{R}{\frac{1 + R}{100}}$$

where, D – oven dry weight
W – weight of water
R – regain
M – moisture content

Correction of yarn count and strength for humidity changes

$$\text{Corrected count} = \frac{N(100 + R_\alpha)}{(100 + R_s)} \text{ (for direct system)}$$

(or)

$$= \frac{N(100 + R_s)}{(100 + R_\alpha)} \text{ (for indirect system)}$$

where N is yarn count at actual regain and R_s is Standard regain; R_α is actual regain

$$\text{Corrected strength} = \frac{S(100 + f\, R_s)}{100 + f\, R_\alpha}$$

where f is correction factor depending on type of fibres and s is strength of yarn.

Calculation of uniformity ratio and percentage of short fibres in a digital fibrograph

$$\text{Percentage of short fibres} = \frac{A_0 - 4A_1 + 3A_2}{A_0}$$

where, A_0 = amount counter reading by depressing 0.15 in. button
A_1 = amount counter reading by depressing 0.45 in. button
A_2 = amount counter reading by depressing 0.55 in. button.

The relationship between specific surface S, the maturity ratio M and the fibre weight per cm H is given by

$$S = \text{constant} \times \frac{1}{\sqrt{MH}}$$

Where, S = Specific surface of fibre
         M = Maturity ratio
         H = Fibre weight/cm

$$\text{Fibre weight/cm, } H = \frac{\text{micronaire estimate of } MH}{\text{test value of } M}$$

Standard fibre weight/cm, $H_s = H/M$

$$\text{Maturity ratio, } M = \frac{\text{micronaire estimate of } MH}{\text{test value of } H}$$

$$\text{Immaturity ratio, } i = \sqrt{0.07D + 1}$$

where, D is average difference between the readings at low compression and high compression.

Fineness in micrograms/in., $W = 4,85,000 \ (I/A_2)$

Where A is specific surface area
Perimeter of fibre in microns, $P = 12,566 \ (I/A)$
Wall thickness 't' in microns $t = 1000 \times T/A$

$$\text{Uniformity ratio} = \frac{50\% \text{ span length}}{2.5\% \text{ span length}} \times 100$$

where $T = \dfrac{2}{1 + \dfrac{\sqrt{1-1}}{1}} \times 100$

Therefore wall thickness of fibre, 't' in microns $t = \dfrac{2000}{A\left[1 + \dfrac{\sqrt{1-1}}{1}\right]}$

Degree of cell wall thickening of fibre, $\emptyset = A/A_1$

Where, A = area of cross section of the cell wall
         A_1 = area of the circle of equal perimeter
         $A_1 = \pi r^2$

$$= \frac{4\pi \times \pi r^2}{4\pi} = \frac{1\pi^2 r^2}{4\pi}$$

Perimeter, $P = 2pr$

Therefore

$$P_2 = (2 \ \pi \ r^2) = 4\pi \ r^2$$

Therefore,

$$A_1 = A_1 = \frac{P_2}{4\pi}$$

Therefore $f = A_1/A = 4 \ \pi \ A/P_2$

Percentage of mature fibres = number of mature fibres/total number of fibres examined × 100

$$\text{Maturity, } M = \frac{N - D + 0.7}{200}$$

Where, N = percentage of normal fibres

D = percentage of dead fibres

The value $N - D$ is known as immaturity count.

$$M = H/H_s$$

where, H = actual fibre weight/cm

H_s = standard fibre weight/cm

$$\text{Maturity coefficient, } M = \frac{N + 0.6H + 0.41}{100}$$

Where, N = percentage of mature fibres

H = percentage of half mature fibres

I = percentage of immature fibres

$$\text{Fibre maturity ratio} = \frac{(\text{Normal fibres} - \text{dead fibres})}{200} + 0.7$$

Fibre quality index, $FQI = l \ usm/f$

Where, lu = product of 2.5% span length (l) and uniformity ratio (u) measured by digital fibrograph, divided by 100

s = bundle fibre strength in g/tex measured on Stelometer at 3 mm gauge length

m = maturity co-efficient

f = fibre fineness, in micrograms per inch, measured on micronaire

In case of a Shirley analyzer, the following are calculated

$$\text{Lint content} = \frac{L_2 + L_4 + L_5}{S} \times 100$$

$$\text{Trash content}, T = \frac{T_5 \times 100}{S}$$

$$\text{Invisible loss} = 100 - (L + T)$$

where,   S is weight of the sample in grams (100 g)

L_2, L_4 and L_5 are lint extracted at second, fourth and fifth time of processing.

T_5 is trash extracted at fifth time of processing.

15.4   Key definitions in mechanical properties of fibres

Load

The application of a lead to a specimen in its axial direction causes a tension to be developed in the specimen. The load is normally expressed in grams weight or pounds weight like gravitational units of force.

Breaking load

It is the load at which the specimen breaks, and it is usually expressed in grams weight or pounds weight.

Stress

It is the ratio between the force applied and the cross section of the specimen.

$$\text{Stress} = \frac{\text{force applied}}{\text{area of cross section}}$$

The force is accurately expressed in dynes or poundals. Hence, the stress on a fibre may be given in terms of dynes per square centimeter (dyn/cm^2).

Specific or mass stress

It is the ratio of the force applied to the linear density (mass per unit length):

$$\text{Specific stress or mass stress} = \frac{\text{force applied}}{\text{mass/unit length}}$$

or $\dfrac{\text{force applied}}{\text{linear density}}$

The mass stress is expressed in grams force weight per denier in the CGS system and millinewtons/tex in the MKS system.

Tenacity or specific strength

It is defined as the mass stress at break and is expressed as grams force per denier in the CGS system and millinewtons/tex in the MKS system.

Breaking length

It is the length of the specimen which will just break under its own weight when hung vertically. The expression of strength in terms of breaking length is useful for comparing the strength of different fibre structures, e.g. for comparing single fibre strength with the yarn strength.

Strain

It is used to relate the stretch or elongation with the initial length. A specimen elongates to a certain length on being subjected to a load. The amount of elongation will depend on the initial length of the specimen.

Strain = elongation/initial length

Extension

It is strain expressed as a percentage

$$\text{Extension} = \dfrac{\text{elongation}}{\text{initial length}} \times 100$$

Breaking extension

It is the extension of the specimen at the breaking point.

Young's Modulus

It is the ratio of stress, strain, and provides a measure of the force required to produce a small extension. A high modulus indicates inextensibility and a low modulus great extensibility. In textiles the term Initial Young's modulus is used and it describes the initial resistance to extension of a textile material.

$$\frac{\text{Area under the curve}}{\text{breaking stress} \times \text{breaking strain}}$$

This would therefore be equal to one half.

For a particular curve this ratio is known as the work factor. Thus, for a curve which is concave towards the strain axis (horizontal), the work factor is greater than one-half; and, conversely if the curve is concave towards the stress axis, the work factor is greater less than one-half.

$$\text{Work factor} = \frac{\text{area under the curve (work of rupture)}}{\text{breaking stress} \times \text{breaking strain}}$$

Elastic recovery

It is the ability of a material to recover from a given extension. It is given as

$$\frac{\text{elastic extension}}{\text{total extension}}$$

Perfectly elastic materials will have an elastic recovery of 1.0, while materials without any power or recovery will have a recovery of zero. Elastic recovery may be expressed as a percentage.

Elastic recovery values are affected by several factors, including the time allowed for recovery, the moisture in the specimen, and the total extension used in the test. Therefore, in order to make comparisons between different materials, it is necessary to specify the conditions under which the elastic recovery is determined.

15.5   Formulae relating to pressley strength tester

(1)   At zero gauge length test,

(a)   Pressley Index, P.I. = $\dfrac{\text{Breaking strength in pounds}}{\text{Bundle weight in milligrams}}$

(b)   Tenacity in g/tex = $5.36 \times$ P.I.

(c)   Tensile strength in 1000 pounds/sq. inch = $10.816 \times$ P.I. $- 0.12$

(2)   At 1/8 inch (3 mm) gauge length test,

(a)   Pressley ratio, P.R. = $\dfrac{\text{Breaking strength in pounds}}{\text{Bundle weight in milligrams}}$

(b)   Tenacity in g/tex = $6.8 \times$ P.R

(c)   Fibre strength index, FSI = P.R $/3.19 \times 100$

(d)   Tensile strength in 1000 pounds/sq. in. = FSI $\times (84/100)$

$$\text{Tenacity in g/tex} = \frac{\text{breaking strength in kg} \times 15 \,(\text{for 3 mm gauge length})}{\text{bundle weight in mg}}$$

$$\text{Tenacity in g/tex} = \frac{\text{breaking strength in kg} \times 11.81 \,(\text{for 0 gauge length})}{\text{bundle weight in mg}}$$

Correction between single strength and CSP

For combed yarns = 0.75 + (0.12 × lea strength in lbs)
For carded yarns   = 0.8 + (0.134 × lea strength in lbs)

15.6   Other formulae

$$\text{Cleaning efficiency of a machine \%} = \frac{T_1 - T_2}{T_1} \times 100$$

where, T_1 = % trash in the material fed
T_2 = % trash in the material delivered

Fibre quality index, $FQI = l\ u\ s\ m\ /f$

where, lu = product of 2.5% span length (l) and uniformity ratio (u) measured by digit fibrograph, divided by 100.
s = bundle fibre strength in g/tex measured on Stelometer at 3 mm gauge length m – maturity co-efficient
f = fibre fineness, in micrograms per inch, measured on micronaire.

$$CSP = (310 - \text{count})\ \sqrt{FQI}$$

In case of a Shirley analyzer, the following are calculated

Lint content = $(L_2 + L_4 + L_5)/S \times 100$
Trash content $T = (T_5 \times 100)/S$
Invisible loss = $100 - (L + T)$

where, S – weight of the sample in grams (100 g)
L_2, L_4 and L_5 – lint extracted at second, fourth and fifth time of processing.
T_5 – trash extracted at fifth time of processing.

16

Calculations in the spinning process

16.1   Introduction

This chapter mainly comprises of material and machine calculations from blow room to spinning. Additionally technical specifications of modern machines are given to enhance the knowledge of the student.

16.2   Basic terminologies

Opening

It consists of mechanically separating the bales into smaller tufts of fibres. This separation releases many of the impurities present in natural fibres and these can be removed as waste.

Cleaning

It is the process of separation of fibres from waste such as dirt, trash etc.

Draft

It is the process of attenuation of the fibres or in other words it is the ratio of the weight per unit length of material fed to that delivered.

Draft constant

It is a constant which is used to find the change wheel required to get the desired draft.

Tension draft

It is a slight tension given to the material. In case of blow room the tension draft is given between the cage delivery roller and calendar roller and between calendar roller and lap roller. The tension draft ranges from 1.03 to 1.05. A higher value will give uneven places and a lower value will make the lap sagging.

Neps

It is a small clump or cluster of shorter fibres.

Noil

These are the shorter fibres that are removed during combing.

Mixing

It is the process of combining identical type of fibres varying in characteristics. Example is mixing of different varieties of cotton varying in staple length, fineness etc.

Blending

It is the process of combining fibres of different types in suitable proportions. Example is blending fibres of cotton and polyester.

16.3   Blow room

The main object of the blow room is to open and clean the material. The opening action enables reduction of the fluff of fibres into smaller tufts and the cleaning action separate the impurities such as trash, dirt etc. from the fibres.

16.3.1 Calculations in opening and cleaning

$$\text{Cleaning efficiency} = \frac{\text{Trash in material fed} - \text{Trash in material delivered}}{\text{Trash in material fed}} \times 100$$

$$\text{Blows per inch of beaters} = \frac{\text{RPM of beater} \times \text{No. of blades}}{\text{Surface speed of feed roller in inches/min}}$$

$$\text{Lint loss in blow room} = \frac{\text{Trash content in cotton} - [(\text{Trash in lap} \times (100 - \text{Total loss})] \times 100}{\text{Total loss}}$$

16.3.2 Lap calculations

Weight of a given length of lap:

Weight in lbs = Length in yard × 840 × Hank

Length of lap fed to card per minute

$$= \frac{\pi \times \text{Diameter of feed roller (inches)} \times \text{Feed Roller RPM}}{36}$$

Time of one lap running on the card

$$= \frac{\text{Length of lap}}{\text{Length of lap fed to card/min}}$$

$$\text{Hank of lap} = \frac{8.33}{\frac{(\text{Lap wt/yd in ozs} \times 7000)}{16}}$$

16.3.3 Draft calculations

Actual draft and mechanical draft of a finisher scutcher

$$\text{Actual draft} = \frac{\text{Weight/Unit length of breaker lap} \times \text{No. of breaker lap doublings}}{\text{Weight/Unit length of finisher lap}}$$

16.3.4 Production in a scutcher

100% per shift production

$$= \frac{\pi \times \text{bottom calender roller diameter (inches)} \times \text{rpm of bottom calender roller} \times 60 \times 8}{36 \times 840 \times \text{hank of lap}}$$

Blow room laps/machine

$$= \frac{1 \times (\text{lap roller rpm} \times \text{lap roll circumference in inches} \times \text{efficiency})}{63 \times \text{hank of lap}}$$

16.3.5 Waste

The blow room waste excluding gutter loss is about the same as the trash in cotton for good cleaning efficiency. The cleaning efficiency (%) and lint in waste (%) in modern blow room lines is given in table below:

Trash in cotton (%)	No. of beating points	Cleaning efficiency (%)
5, 0 & above	5	60–65
3,0 to 4.9	4	55–60
1.0 to 2.9	3	50–55

Gutter loss will vary between 0.5% (fine mixing) to 1.0% (coarser mixings). The lint in waste of the individual beating points will be in the range of 20 to 30%.

Mixing bale opener	400 kg/h (with opener and cleaner)
	600 kg/h (without opener and cleaner)
Mono cylinder	500 kg/h
Uni mix	400 kg/h
ERM cleaner	400 kg/h
Aero feed	80 kg/h/chute
	400 kg per feeding machine
Scutcher	350–400 kg/h for two scutchers

16.3.6 Production data (Reiter line)

Mixing bale opener	400 kg/h (with opener and cleaner)
	600 kg/h (without opener and cleaner)
Mono cylinder	500 kg/h
Uni mix	400 kg/h
ERM cleaner	400 kg/h
Aero feed	80 kg/h/chute
	400 kg per feeding machine
Scutcher	350–400 kg/h for two scutchers

16.3.7 Technical specifications of modern blow room line

(A) *Bale plucker or super blender*

Plucker in carriages 2 sizes	1700 and 2250 mm
Length of machine carriage	15000 to 45000 mm
Width of machine	3145 and 6200 mm double blade
Weight of the machine	1700 mm – 3500 kg; 2250 mm – 3700 kg
Beater speed	1400 rpm
Production	For 1700 mm – 800 kg; for 2250 mm – 1100 kg.
Power required	5.18 for 1700 mm; 6.18 for 2250

(B) *Blending feeder*

Length	7380 mm; 9380 mm; 11380 mm
Weight	2813 kg; 3045 kg; 3300 kg
Width: Working width	1000 mm; Frame width 1500 mm
Production per hour	200 kg
Main motor	2.2 k W
Beater motor	1.5 kW

(C) *Transport lattice*

Length	6020 mm to 14020 mm (depending on sections)
Frame width	520 mm; working width 420 mm.
Conveyor speed	75 m/mm.
Geared motor	0.75 kW or 1.0 kW

(D) *Two beater opener*

Length	1360 mm
Frame width	1370 mm
Working width	1280 mm
Production	800 kg/h
Beater diameter	610 mm
Motor	2.2 kW
Weight	1090 kg

(E) *Cage condenser*

Air capacity	3500 m³/hour to 4500 m³/hour
motor for cage	0.75 kW
fan motor	3 kW
weight	300 to 450 kg

(F) *Step cleaner (six cylinders)*

Length	3320 mm
Width	1670 mm frame; 1000 mm – working
Production	600 kg/h
Main motor	2.2 kW
Beater motor	1.1 kW
Geared motor	0.25 kW
Weight	2420 kg

(G) *Horizontal opener (porcupine or clothed type)*

Length	1500 mm
Width	1700 mm frame; 1200 mm – working
Production	600 kg/h
Main motor	3 kW
Geared motor	1.1 kW
Weight	1850 kg

(H) *Dust separator*

Length	1500 mm
Width	1700 mm frame; 1200 mm – working
Production	600 kg/h
Main motor	3 kW
Geared motor	1.1 kW
Weight	1850 kg

(I) *Auto mixer*

Length	2500 mm
Width	1500 mm – frame; 1200 mm – working
Production	20–40 kg/cell
Cage motor and lattice drive	1.1 kW
Opening roll	2.1 kW
Geared motor	0.15 kW conveyor drive
Motor fan	4.15 kW
Air capacity	4500-7500 m3/h
Weight	3500 kg

(J) *Card chute feeding system*

Motor fan	4 kW
Fan diameter	425 mm
Speed	2500 mm
Air capacity	5000 cubic m/h
Working width	900 mm
Geared motor for feed	0.25 kW
Blowing motor fan	0.75 kW
Opening roller drive	0.75 kW
Weight	630 kg

16.3.8 Gearing calculations in blow room line

Step cleaner

$$\text{Speed of beater} = \frac{\text{motor speed} \times \text{motor pulley diameter}}{\text{machine pulley diameter}}$$

(i) Speed of beater = $1430 \times 4.33/11.81 = 524.29$ rpm
(ii) Speed of pre-opener = $1410 \times 3.35/9.85 = 479.54$ rpm
(iii) Speed of feed rollers = $1395 \times 1.37/ 4.72 \times 1/10 \times 15/70 \times 55/55$
   = 8.67 rpm

Mixing bale opener

(i) Speed of beater = $950 \times 4.7/5.5 = 811.8$ rpm
(ii) Speed of feed roller = $105 \times 7.5/10 \times 5/11.81 \times 10/10$
   = 33.34 rpm
(iii) Speed of stripper = $950 \times 4.7/5.5 \times 5/15.5 = 261.87$ rpm
(iv) Speed of evener roller = $950 \times 4.7/5.5 \times 5/15.5 \times 14/7.2$
   = 509.20 rpm
(v) Speed of inclined
   spiked lattice = $105 \times 7.5/10 = 78.75$ rpm
(vi) Speed of feed apron (1) = $105 \times 7.5/10 \times 16/53 \times 20/52 = 9.14$ rpm
(vii) Speed of feed apron (2) = $1395 \times 21/70 \times 21/72 = 122$ rpm

Step Cleaner-Gearing Plan

16.1 Gearing plan of a step cleaner

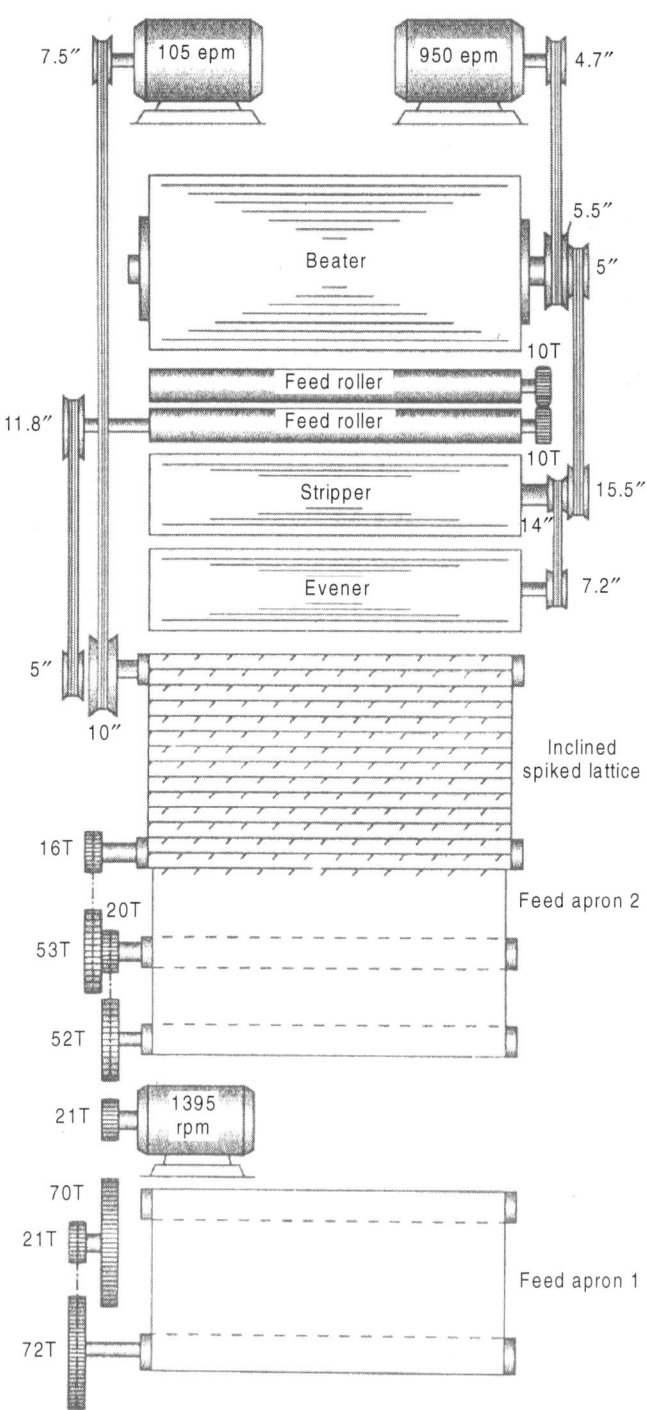

16.2 Gearing plan of mixing bale opener

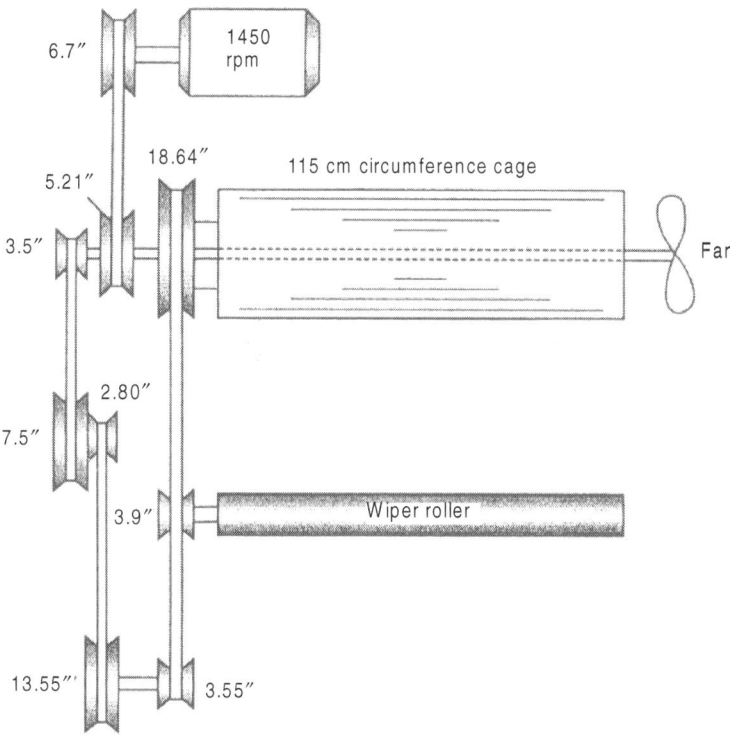

Figure 3.3   Gearing plan of a condensor cage

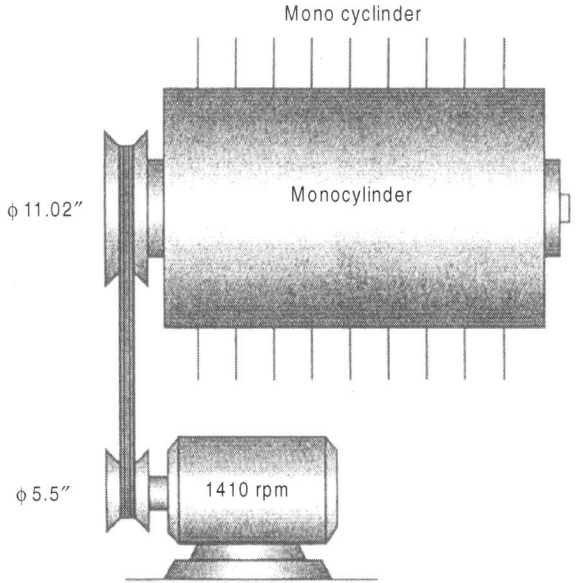

16.3(a) Gearing plan of monocylinder.

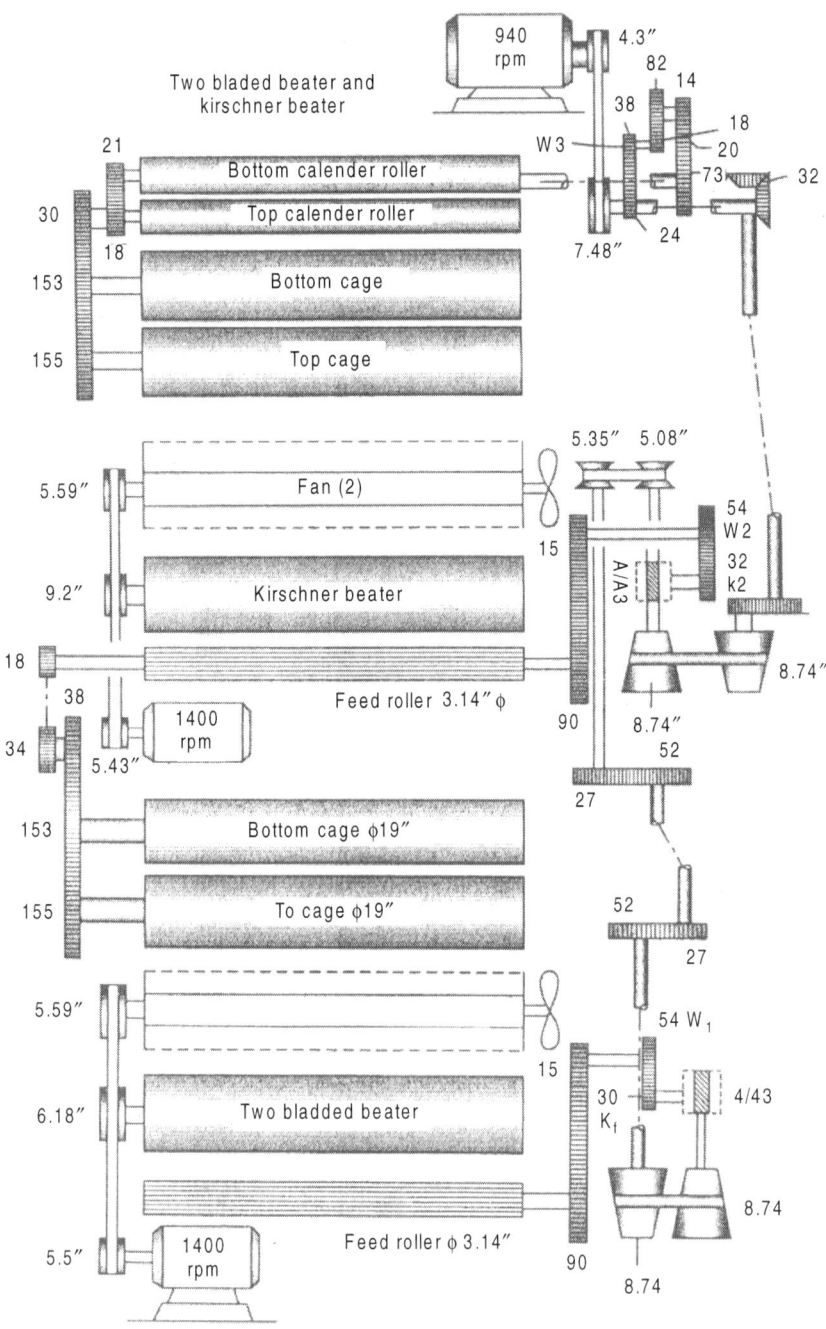

(a) W_1, K_1 are feed change gear for two bladder beater
(b) W_2, K_2 are feed change gear for Kirschner beater
(c) W_3-Delivery change gear

16.3(b) Gearing plan of two bladed beater and Kirschner beater.

Mono cylinder

Speed of mono cylinder = 1410 × 5.5/11.02

Condenser

(i)   speed of fan   = 1450 × 6.7/5.21 = 1864.68 rpm
(ii) speed of cage   = 1450 × 6.7/5.21 × 3.5/7.5 × 2.86/13.55 ×
                              3.55/18.64 = 34.98 rpm
(iii) speed of wiper roller = 1450 × 6.7/5.21 × 3.5/7.5 × 2.86/13.55 ×
                              3.55/3.9 = 167.18 rpm

Two bladed and kirschner beater

Two bladed beater

$$\text{Beats/inch} = \frac{\text{beater speed} \times \text{no. of blades}}{\text{Surface speed of feed roller in inches/mm}}$$

(i)   Speed of two bladed beater = 1400 × 5.5/6.18 = 1245.9 rpm
(ii) Speed of fan (1)   = 1400 × 5.5/5.59 = 1377.45 rpm
(iii) Speed of feed roller   = 940 × 4.3/7.48 × 32/32 × 52/27 ×
                                  8.74/8.74 × 5.08/5.35 × 27/52 × 52/
                                  27 × 8.74/8.74 × 4/43 × 30/56 ×15/
                                  90 = 8.3 rpm
(iv) Surface speed of feed roller = πDN
                                   = π × 3.14 × 8.3
                                   = 81.876 inches/mm.
(v)   Beats/inch for two bladed beater

$$= \frac{\text{beater speed} \times \text{no. of blades}}{\text{surface speed of feed roller}}$$

$$= \frac{1245.9 \times 2}{81.876} = 30.43$$

Kirschner beater

(i)   Speed of kirschner beater = 1400 × 5.43/9.2 = 826.30 rpm
(ii) Speed of fan (2)   = 1400 × 5.43/5.59 = 1359.92 rpm
(iii) Speed of feed roller   = 940 × 4.3/7.48 × 32/32 × 52/21 ×
                                  8.74/8.74 × 4/43 × 52/54 × 15/90 =
                                  9.765 rpm.
(iv) Surface speed of feed roller = πDN = π × 3.14 × 9.665 = 96.327
                                   inches/min

Cage

Cage speed for kirschner bladed beater
= 940 × 4.33/7.48 × 24/38 × 18/82 × 14/73 × 21/18 × 30/155
= 3.3 rpm
Cage speed for two bladed beater
= 8.365 × 18/34 × 38/155 = 1.08 rpm

Hopper feeder and scutcher

(i) Speed of stripper roller   = 1430 × 1.6/8.66 = 264.2 rpm
(ii) Speed of bottom stripper roller = 1430 × 1/6.015 = 380.38 rpm
(iii) Speed of regulating roller   = 1430 × 1.6/11.02 = 207.62 rpm
(iv) Speed of clearer roller   = 1430 × 1.4/11.02 × 7.64/5.76
                                     = 275.38 rpm
(v) Speed of feed roller   = 1380 × 28/70 × 26/72 × 1/10
                                     = 19.93 rpm
(vi) Speed of feed apron   = 940 × 4.3/2.48 × 32/32 × 52/27 ×
                                     8.74/8.74 × 5.08/5.35 × 27/52 ×
                                     52/27 × 8.74/8.74 × 4/43 × 40/50
                                     × 16/53 × 20/53 = 8.5 rpm
(vii) Speed of inclined spiked lattice = 940 × 4.3/2.48 × 32/32 × 52/57 ×
                                     8.74/8.74 ×5.08/5.35 × 27/52 × 52/
                                     27 × 8.74/8.74 × 4/43 × 40/50
                                     = 74.37 rpm
(viii) Speed of feed roller 1   = 940 × 4.3/7.48 × 32/32 × 52/27 ×
                                     8.74/8.74 × 5.08/5.35 × 27/52 ×
                                     52/27 × 5.74/8.74 × 4/43 × 30/56
                                     × 15/90 = 8.3 rpm
(ix) Speed of adjustable lattice 1   = 8.3 × 18/29 = 5.15 rpm
(x) Speed of adjustable lattice 2   = 8.3 × 18/28 = 5.335 rpm
(xi) Speed of calendar roller   = 940 × 115/190 × 24/38 × 18/52 ×
       (bottom)   14/23 = 15.12 rpm
(xii) speed of second calendar roller = 940 × 115/190 × 24/38 × 18/82 ×
                                     14/73 × 21/18 = 17.64 rpm
(xiii) Speed of third calendar roller = 17.64 × 18/19 = 16.71 rpm
       (128 Ø)
(xiv) Speed of top calendar roller   = 16.71 × 19/28 = 11.33 rpm
       (148 Ø)
(xv) Speed of shell rollers   = 960 × 120/210 × 40/157 × 21/44
                                     = 66.7 rpm
(xvi) Surface speed of shell roller = $\pi DN = \pi × 66.7 × 2.45/25.4$
                                     = 2021.19 inches/min.

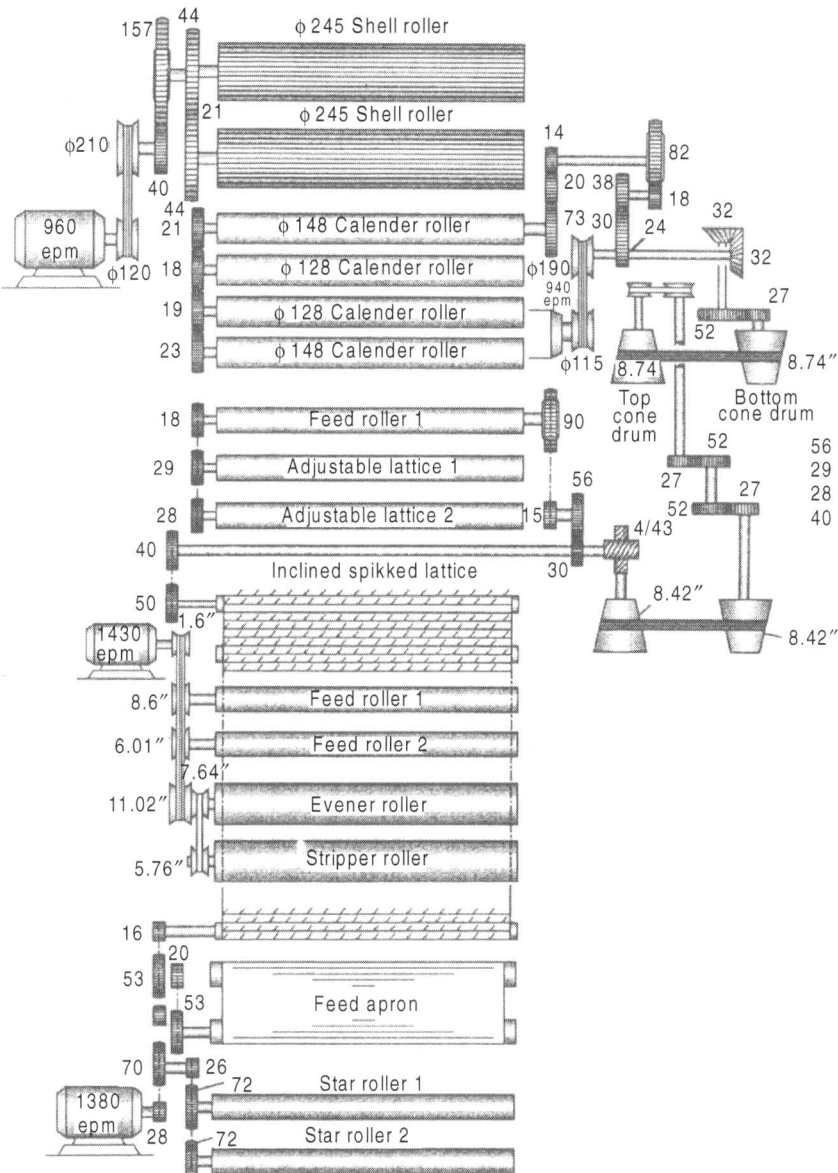

16.4 Gearing plan of a hopper feeder and scutcher.

16.4   Carding

The main object of the carding process is to individualize the fibres and convert the fibres into a strand called the sliver.

16.4.1 Miscellaneous formulae

Length of Sliver produced by a card

= Length of lap fed × draft in card

Weight of card sliver delivered

$$= \frac{\text{Weight of lap fed} \times (100 - \text{loss}\%)}{100}$$

Weight/yard of card sliver

$$= \frac{\text{Weight of lap fed} \times (100 - \text{Waste}\,\%) \times 7000}{1 \times 100 \times (\text{draft} \times \text{length of lap}) \times 1}$$

Card doffer speed

$$= \frac{0.036 \times \text{Cylinder rpm}}{\text{Doffer diameter} \times \text{staple length}}$$

Diameter of sliver trumpet bore

$$= 0.22 \sqrt{\text{One yard sliver weight in grains}}$$

Card production/8 hours

$$= \frac{1.484 \times \text{Doffer rpm} \times \text{efficiency in lbs}}{\text{Hank of sliver}}$$

Carding quality index $= \dfrac{C + N + L + E}{4}$

where C = Cleaning efficiency Index
   N = Nep Index
   L = Length Index
   E = Evenness Index

Carding index

$$= \frac{\text{Cylinder rpm} \times \text{surface area of cylinder in meters}}{\text{Taker in speed in meter/min} \times \text{weight of one meter lap in gm}}$$

Actual draft $= \dfrac{\text{Weight per yard fed}}{\text{Weight per yard delivered}}$

Sliver weight per yard in grains

$$= \frac{\text{Lap weight per yard in grains}}{\text{Actual Draft}}$$

Mechanical draft

$$= \frac{\text{Weight of one yard of lap}}{\text{Weight of Sliver}}$$

Weight of one yard lap

$$= \frac{(100 - \text{Waste \%}) \times \text{Weight of sliver} \times \text{Actual draft}}{100}$$

Total draft in a draw frame

$$= \frac{\text{Hank delivered} \times \text{ends up}}{\text{Hank fed}}$$

16.4.2 Waste calculations

Lickerin waste %

$$= \frac{\text{Weight in grams of licker in droppings}}{\text{Weight of lap in grams}} \times 100$$

Fan waste %

$$= \frac{\text{Weight of fan waste in grams}}{\text{Weight of lap in grams}} \times 100$$

Cylinder and doffer fly waste %

$$= \frac{\text{Weight of cylinder and doffer fly waste in grams}}{\text{Lap weight in grams}} \times 100$$

Flat waste %

$$= \frac{\text{weight of flat waste in grams}}{\text{Weight of lap in grams}} \times 100$$

Total waste in a card
= Lickerin waste + fan waste + cylinder and doffer fly + flat waste

Standard nep count in card
= 0.2 × (neps/board × hank card × card width in cm)

$$\text{Neps/100 mg} = \frac{\text{Standard nep count}}{3.1}$$

The table below gives the standard waste (%) in cards

Mixing	Type of cards	Licker-in droppings	Flat strips	Others	Total
20s to 40s	SHP	1.5	2.5	0.5	4.5
	HP	1.8	3	1.2	6
	VHP	2	3.5	1.5	7. 0
50s to 100s	SHP	1.2	2.3	0.5	4
	HP	1.5	2.8	1.2	5.5
	VHP	1.8	3	1.2	6

16.4.3 Technical specifications of a high production card

Maximum fibre length	Up to 60 mm – cotton, synthetic fibres and blends
Maximum mechanical production	100 kg/h
Batt weight at inlet	300 g/m to 1000 g/m
Delivery sliver count	3.3 to 15 g/m or 0.04 to 0.18 Ne
Total draft	57 to 304
Delivery draft	1 to 1.5
Total waste	from 2 to 6%
Working width	1016 mm or 40"
Feed roller diameter	80 mm
Licker in diameter	350 mm clothed
Licker in speed	348 to 1393 rpm
Bottom cylinder diameter	256 mm
Bottom cylinder speed	14.5 to 29 rpm
Carding cylinder diameter	1290 mm clothed
Carding cylinder speed	300–600 rpm
No. of working flats	38
Total flats	100
Stationary flats at inlet	2; optional 5
Stationary flats at delivery	3; optional 5 + husk knife or 6
Doffer diameter	760 mm; speed – 7 to 79 rpm
Web detaching cylinder	Diameter 90 mm; speed – 72 to 793 rpm
Delivery draft group	2 over 2
Draft roller diameter	40 mm
Main motor	4 kW
Inlet and delivery drive motor	3 kW
Flats drive motor	0.185 kW
Brush drive motor	0.185 kW
Auto leveler drive motor	0.36 kW
Suction fan motor	2.2 kW

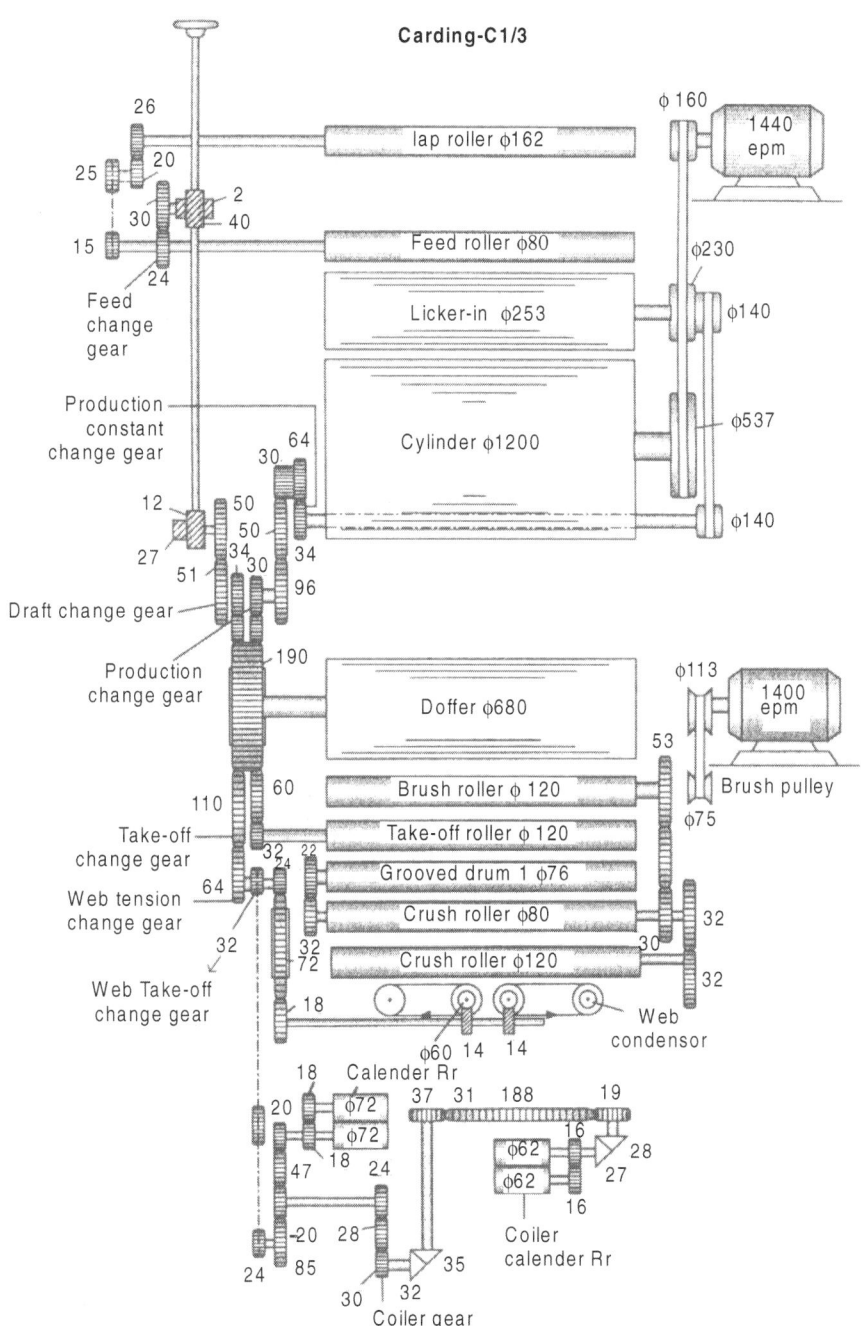

Carding-C1/3

26
φ 160
1440 epm
25
20
lap roller φ162
2
30   40
15
Feed roller φ80
φ230
24
Feed change gear
Licker-in φ253
φ140
Production constant change gear
64
30
Cylinder φ1200
φ537
12   50
50
φ140
27   34   30   34
51   96
Draft change gear
Production change gear   190
φ113
1400 epm
Doffer φ680
53
60   Brush roller φ 120   Brush pulley
110
Take-off change gear
32   Take-off roller φ 120   φ75
24   22
Web tension change gear   64   Grooved drum 1 φ76
32   Crush roller φ80   32
32   Crush roller φ120   30
72
Web Take-off change gear   18   Web condensor
φ60 14   14
18   Calender Rr
20   φ72   37   31   188   19
φ72   16
47   18   24   φ62   28
φ62   27
16
-20   28   Coiler calender Rr
24   85   35
30   32
Coiler gear

16.5 Gearing plan of a carding machine.

16.4.4 Gearing calculations in carding

Speed calculations

(i) Speed of lap roller

$$= 1440 \times \frac{6.3}{9.05} \times \frac{5.5}{5.5} \times \frac{26}{70} \times \frac{30}{96} \times \frac{28}{32} \times \frac{50}{31} \times \frac{12}{27} \times \frac{2}{40} \times \frac{30}{24} \times \frac{15}{25} \times \frac{20}{26}$$

= 1.3 rpm

(ii) Speed of feed roller

$$= 1440 \times \frac{6.3}{9.05} \times \frac{5.5}{5.5} \times \frac{26}{70} \times \frac{30}{96} \times \frac{28}{32} \times \frac{50}{31} \times \frac{12}{27} \times \frac{2}{40} \times \frac{30}{24}$$

= 2.7 rpm

(iii) Speed of licker in $= 1440 \times \dfrac{6.3}{9.05} = 1002.4$ rpm

(iv) Speed of cylinder $= 1440 \times \dfrac{6.3}{21.14} = 429.13$ rpm

(v) Speed of doffer $= 1440 \times \dfrac{6.3}{9.05} \times \dfrac{5.5}{5.5} \times \dfrac{26}{70} \times \dfrac{30}{96} \times \dfrac{20}{190} = 17.15$ rpm

(vi) Speed of brush roller $= 1440 \times \dfrac{130}{65} = 2880$ rpm

(vii) Speed of take up roller

$$= 1440 \times \frac{6.3}{9.05} \times \frac{5.5}{5.5} \times \frac{26}{70} \times \frac{30}{96} \times \frac{28}{190} \times \frac{190}{32} = 101.8 \text{ rpm}$$

(viii) Speed of delivery calendar roller

$$= 1440 \times \frac{6.3}{9.05} \times \frac{5.5}{5.5} \times \frac{26}{70} \times \frac{30}{96} \times \frac{28}{190} \times \frac{190}{35} \times \frac{32}{64} \times \frac{35}{20} \times \frac{16}{16}$$

= 217.19 rpm

(ix) Speed of coiler calendar roller

$$= 17.15 \times \frac{190}{35} \times \frac{32}{24} \times \frac{35}{20} \times \frac{24}{16} \times \frac{32}{24} \times \frac{37}{20} \times \frac{28}{16} = 320.87 \text{ rpm}$$

(x) Surface speed of feed roller

$$= \pi \, DN = \pi \times \frac{80}{25.4} \times 2.7 = 26.72 \text{ inches/min}$$

(xi) Surface speed of calendar roller

$$= \pi \, DN = \frac{72}{25.4} \times 217.19 = 1934.14 \text{ inches/mm}$$

Draft calculations

(i) Draft constant $= \dfrac{\text{Mechanical draft}}{\text{Draft change wheel}}$

(ii) Mechanical draft $= \dfrac{\text{Surface speed of calendar roller}}{\text{Surface speed of feed roller}} = \dfrac{1934.14}{26.71} = 73$

(iii) Draft constant $= \dfrac{\text{Surface speed of coiler calendar roller (DCW)}}{\text{Surface speed of lap roller}}$

$$= \left(\pi \times 62 \times 1 \times \frac{26}{20} \times \frac{25}{15} \times \frac{24}{30} \times \frac{40}{2} \times \frac{27}{12} \times \right.$$
$$\frac{51}{50} \times \frac{35}{36} \times \frac{1}{35} \times \frac{32}{24} \times \frac{35}{20} \times \frac{24}{30} \times \frac{32}{25} \times$$
$$\left. \frac{37}{188} \times \frac{167}{27} \times \frac{28}{27} \right) \Big/ \pi \times 162 \times 1 = 2.6$$

Draft checking

Draft constant $= \dfrac{\text{Mechanical draft}}{\text{DCW}} = \dfrac{73}{28} = 2.6$

(iv) Draft constant $= \dfrac{\text{Surface speed of coiler calendar roller}}{\text{Surface speed of doffer roller}}$

$$= \frac{\pi \times 62 \times 320.87}{\pi \times 680 \times 17.15}$$

Production calculation

Assume that the hank delivered = 0.14
Efficiency = 85%
Production in kg/shift

$$= \pi \times \frac{680 \times 17.15 \times 60 \times 8 \times 85 \times 1.7 \times 35 \times 840 \times 0.14 \times 2.205 \times 100}{25.4}$$

= 107.04 kg

16.5   Draw frames

The object of the drawing process is to straighten the fibres by drafting and reduce the unevenness of the sliver by doubling a group of card slivers.

16.5.1 Gearing calculations of draw frame

Speed calculations

(i) Speed of front roller $= 1440 \times \dfrac{245}{165} \times \dfrac{33}{84} \times \dfrac{84}{33} = 2138.18$ rpm

(ii) Speed of second roller $= 1440 \times \dfrac{245}{165} \times \dfrac{33}{84} \times \dfrac{84}{33} \times \dfrac{35}{25} = 2993.45$ rpm

(iii) Speed of third roller $= 1440 \times \dfrac{245}{165} \times \dfrac{33}{131} \times \dfrac{42}{69} \times \dfrac{24}{144} \times \dfrac{143}{18} \times \dfrac{28}{20}$

$= 607.753$ rpm

(iv) Speed of fourth roller $= 1440 \times \dfrac{245}{165} \times \dfrac{33}{131} \times \dfrac{42}{69} \times \dfrac{24}{144} \times \dfrac{143}{18}$

$= 434.10$ rpm

(v) Speed of back roller $= 1440 \times 1440 \times \dfrac{245}{165} \times \dfrac{33}{131} \times \dfrac{42}{69}$

$= 327.85$ rpm

Draft calculations

(i) Front zone draft $= \dfrac{\text{Surface speed of second roller}}{\text{Surface speed of third roller}}$

$= \dfrac{\pi \times 25 \times 2993.45}{\pi \times 20 \times 607.753} = 6.156$

(ii) Back zone draft (break draft) $= \dfrac{\text{Surface speed of fourth roller}}{\text{Surface speed of the back roller}}$

$= \dfrac{\pi \times 28 \times 434.10}{\pi \times 28 \times 327.15} = 1.324$

(iii) Total draft (mechanical draft) $= \dfrac{\text{Surface speed of second roller}}{\text{Surface speed of back roller}}$

$= \dfrac{\pi \times 25 \times 2993.45}{\pi \times 28 \times 327.15} = 8.152$

Total draft = Front zone draft × Break draft
$= 6.156 \times 1.324$
$= 8.150$

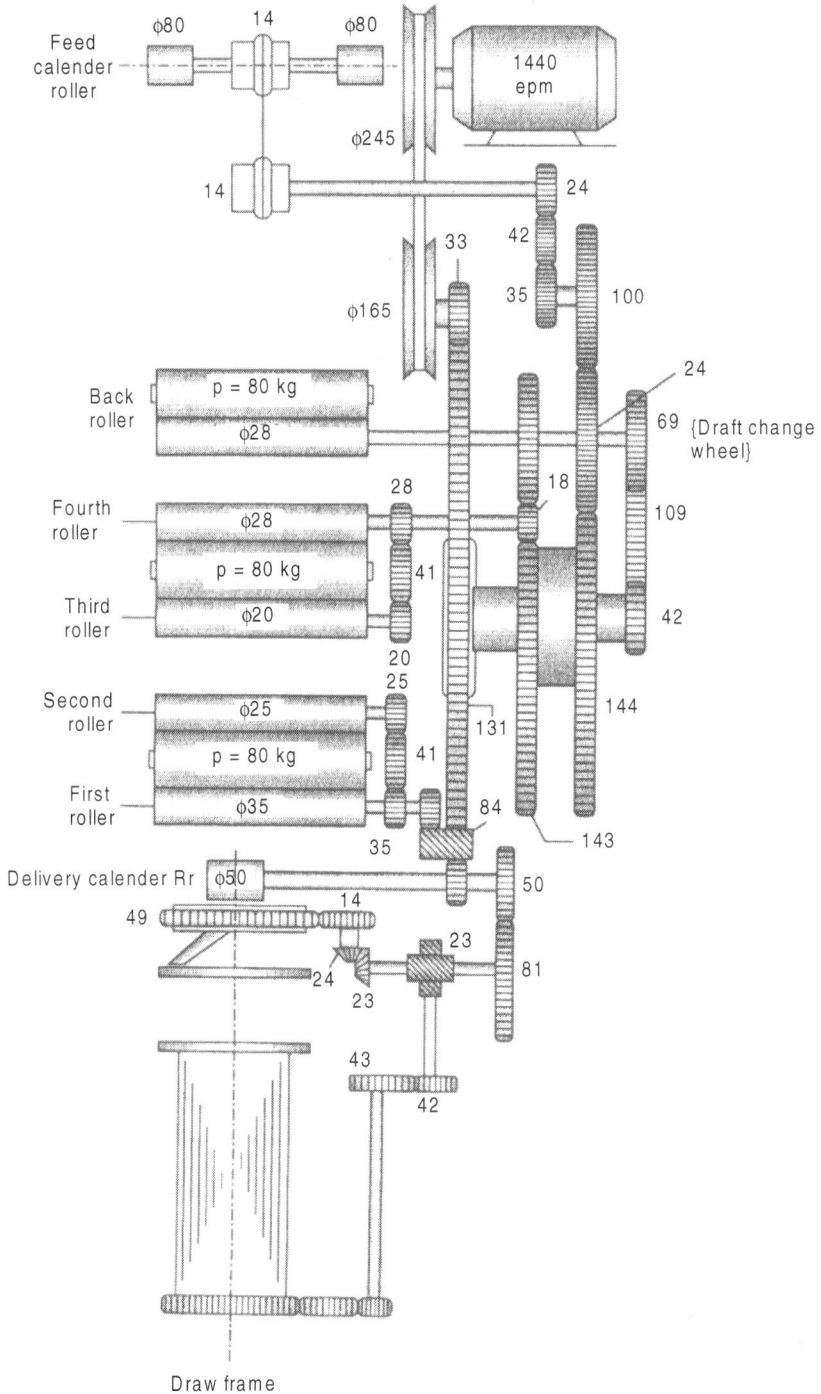

16.6 Gearing plan of a draw plan.

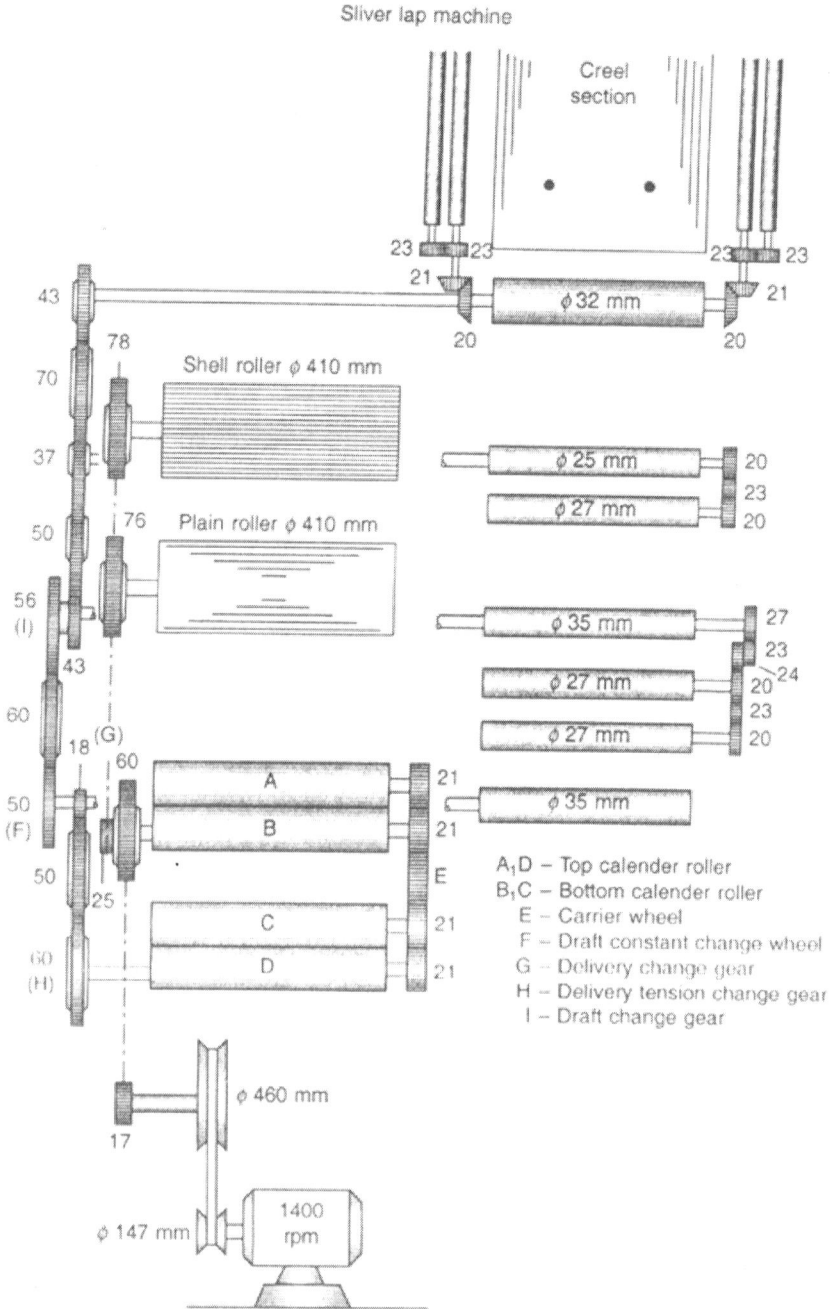

16.6(a) Gearing plan of a silver lap machine.

(iv) Draft constant $= \dfrac{\text{Surface speed of front roller}}{\text{Surface speed of back roller}}$

When DCW= 1
Back roller speed = 1

$$= \pi \times 35 \times \dfrac{\dfrac{\text{DCW}}{42} \times \dfrac{131}{84} \times \dfrac{84}{33}}{\pi \times 28 \times 1}$$

$$= \dfrac{\pi \times 35 \times 1 \times 131 \times 84}{\pi \times 28 \times 1 \times 42 \times 84 \times 33}$$

Therefore draft constant $= 0.1181$

Draft checking:

Draft = draft constant × draft change wheel
   $= 0.1181 \times 69$
   $= 8.1489$

(v) Web tension draft $\quad = \dfrac{\text{Surface speed of calendar roller}}{\text{Surface speed of front roller}}$

Speed of calendar roller $\quad = 1440 \times \dfrac{245}{165} \times \dfrac{33}{47} = 1501.27$ rpm

Therefore web tension draft $= \dfrac{\pi \times 50 \times 1501.27}{\pi \times 35 \times 2138.18} = 1.003$

(vi) Creel tension draft $\quad = \dfrac{\text{Surface speed of back roller}}{\text{Surface speed of creel calendar roller}}$

Speed of calendar roller $\quad = 1440 \times \dfrac{245}{165} \times \dfrac{33}{131} \times \dfrac{42}{69} \times \dfrac{24}{144} \times$
$\dfrac{144}{100} \times \dfrac{35}{24} \times \dfrac{14}{14}$

$= 114.75$ rpm

Thus, creel tension draft $\quad = \dfrac{\pi \times 28 \times 327.85}{\pi \times 80 \times 114.75}$
$= 0.999$

Production calculations

Required data:
   Front roller speed $\quad\quad = 2138.18$ rpm
   Front roller diameter 35 mm $= 1.377$ in.

Hank of the sliver   = 0.135
Shift of 8 hours
No. of deliveries   = 2
Efficiency   = 85%

Production/shift of 8 hours = $\dfrac{\pi DN \times 60 \times 8 \times \text{efficiency} \times \text{no. of deliveries}}{36 \times 840 \times 2.204 \times \text{hank} \times 100}$

$$= \dfrac{\pi \times 1.377 \times 2138.18 \times 60 \times 85 \times 2}{36 \times 840 \times 2.204 \times 0.1353 \times 100}$$

$$= 838.86 \text{ kg/2 deliveries}$$

$$= 419.43 \text{ kg/delivery}$$

16.6   Sliver lap machine

The importance of lap preparation is as follows:

(i)   Fibres from carded sliver are not well oriented and may lie in all directions.
(ii) Most of the fibres arc hooked.
(iii) To avoid excessive strain on comber needles.
(iv) To avoid loss of good fibres in the nods.
(v)   To get high degree of evenness in transverse direction.

16.6.1 Miscellaneous formulae

$$\text{Silver weight/yard in a silver lap} = \frac{\text{Weight of silver lap}}{\text{No. of silver ends fed}}$$

$$\text{Creel silver hank} = \frac{8.33}{\text{creel silver weight/yard}}$$

Silver laps/Machine

$$= \frac{1 \times (\text{lap drum rpm} \times \text{drum circumference in inches} \times \text{efficiency})}{63 \times \text{hank of lap}}$$

16.6.2 Gearing calculations of a sliver lap machine

Speed calculations

(i) Speed of calendar roller = $1400 \times \dfrac{147}{460} \times \dfrac{17}{60} = 126.76$ rpm

(ii)   Speed of front roller $= 1400 \times \dfrac{147}{460} \times \dfrac{17}{60} \times 60 = 422.53$ rpm

(iii) Speed of fourth roller $= 422.53 \times \dfrac{50}{56} = 377.26$ rpm

(iv)   Speed of third roller $= 377.26 \times \dfrac{27}{23} \times \dfrac{24}{20} = 531.44$ rpm

(v)   Speed of second roller $= 331.44$ rpm

(vi)   Speed of sixth roller $= 377.26 \times \dfrac{43}{37} = 438.43$ rpm

(vii) Speed of fifth roller $= 438.43$ rpm.
(viii) Speed of feed roller $= 377.26$ rpm

(ix)   Speed of shell roller $= 1400 \times \dfrac{147}{460} \times \dfrac{17}{60} \times \dfrac{25}{78} = 40.628$ rpm

(x)   Speed of plain drum $= 1400 \times \dfrac{147}{460} \times \dfrac{17}{60} \times \dfrac{25}{76} = 41.607$ rpm

Draft calculations

(i)   Actual draft $= \dfrac{\text{Surface speed of shell roller}}{\text{Surface speed of feed roller}}$

$= \dfrac{\pi \times 35 \times 422.53}{\pi \times 27 \times 438.43} = 1.2492$

(ii)   Front zone draft $= \dfrac{\text{Surface speed of first roller}}{\text{Surface speed of second roller}}$

$= \dfrac{35 \times 422.53}{27 \times 531.44} = 1.0306$

(iii) Back zone draft $= \dfrac{\text{Surface speed of fifth roller}}{\text{Surface speed of sixth roller}}$

$= \dfrac{\pi \times 35 \times 377.26}{\pi \times 28 \times 438.43} = 1.1154$

(iv)   Middle zone draft $= \dfrac{\text{Surface speed of third roller}}{\text{Surface speed of fourth roller}}$

$= \dfrac{27 \times 531.44}{35 \times 377.26} = 1.0866$

(v) Total draft = Front zone draft × Middle zone draft × Back zone draft
= 1.0306 × 1.1154 × 1.0866 = 1.2492

To find draft constant

Assume the delivery speed to be 1 rpm and draft change wheel (DCW) is 1.

$$\text{Then, Draft constant} = \frac{\pi \times 35 \times 1}{\pi \times 27 \times 50/\text{DCW} \times 43/37} \times 1$$

$$\text{Draft} = \text{Draft constant} \times \text{DCW}$$
$$= 0.02230 \times 56 = 1.2492$$

Production calculations

Let the feed hank = 0.13

$$\text{Actual hank} = \frac{\text{Delivered hank}}{\text{Fed hank}} \times \text{No. of doublings}$$

$$1.37 = \frac{\text{Delivered hank}}{0.13} \times 16$$

Delivery hank = 0.01113

$$\text{Production per shift} = \frac{\pi \, \text{DN} \times 60 \times 8 \times \text{efficiency\%}}{840 \times 2.204 \times 36 \times \text{hank}}$$

$$= \frac{\pi \times 16.14 \times 40.268 \times 60 \times 8 \times 85}{2.204 \times 840 \times 36 \times 0.01113}$$

$$= 1126.04 \text{ kg/shift.}$$

16.6.3 Technical specifications of lap former and ribbon lap machine lap former

No. of ends up	24 or 36
Feed sliver weight	0.12 to 0.18 for 24 slivers
Weight of feed sliver	75 g/m to 120 g/m
Draft	1.5 to 2.0
Lap weight	50–70 g/m
Total lap weight	12 kg
Working speed	Up to 65 m/mm
No. of heads	6
No. of doublings	6
Width of lap	230 mm or 250 mm
Weight/m fed	44 to 75 g/m

Mechanical draft	4–9
Width of lap delivered	265 mm or 300 mm
Weight/m delivered	50 g/m
Total lap weight	13 kg or 15 kg
Delivery speed	Up to 65 m/min

16.7   Ribbon lap machine

16.7.1 Gearing calculations for a ribbon lap machine

Speed calculations

(i)   Speed of feed rollers $= 950 \times \dfrac{147}{280} \times \dfrac{30}{120} \times \dfrac{40}{56} \times \dfrac{47}{62} \times \dfrac{12}{20}$

$= 40.509$ rpm

(ii)   Speed of back roller $= 950 \times \dfrac{147}{280} \times \dfrac{30}{120} \times \dfrac{40}{56}$

$= 89.0625$ rpm

(iii) Back roller speed $= \dfrac{27}{40} \times \dfrac{35}{18} = 116.89$ rpm

(iv)   Speed of front roller $= 950 \times \dfrac{147}{280} = 498.75$ rpm

(v)   Speed of table calendar roller $= 950 \times \dfrac{147}{280} \times \dfrac{35}{81} \times \dfrac{21}{43} \times \dfrac{19}{21} \times \dfrac{20}{59}$

$= 226.28$ rpm

(vi)   Speed of the calendar rollers $= 950 \times \dfrac{147}{280} \times \dfrac{35}{81} \times \dfrac{82}{20}$

$= 120.04$ rpm

(vii) Speed of plain roller drum $=$ Speed of calendar rollers $\times \dfrac{25}{76}$

$= 41.46$ rpm

(viii) Speed of shell roller $=$ speed of calendar rollers $\times \dfrac{25}{77}$

$= 40.92$ rpm

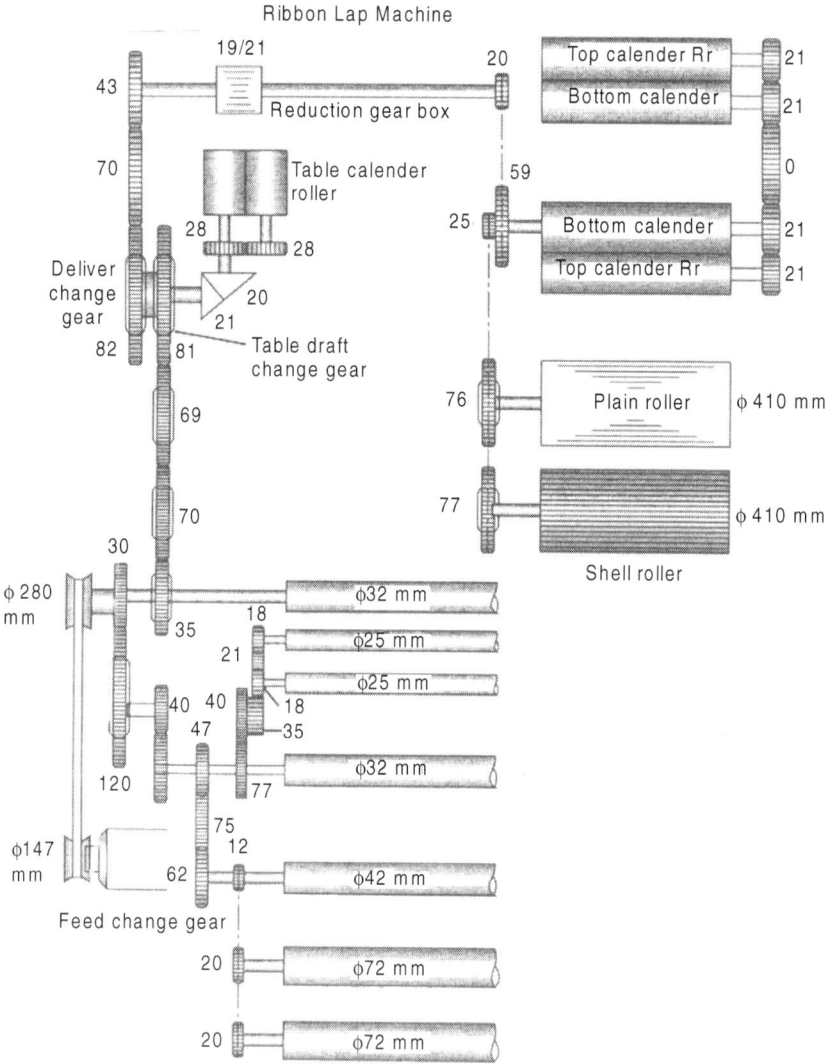

16.7 Gearing plan of a ribbon lap machine.

Draft calculations

(i) Total draft = $\dfrac{\text{Surface speed of delivery roller}}{\text{Surface speed of feed roller}}$

$= \dfrac{\pi \times 410 \times 40.92}{\pi \times 72 \times 40.50} = 5.75$

(ii) Total draft in drafting zone $= \dfrac{\text{Surface speed of front roller}}{\text{Surface speed of back roller}}$

$$= \frac{\pi \times 32 \times 498.75}{\pi \times 30 \times 89.0625} = 5.6$$

To find the draft constant:

$$\text{Draft} = \frac{\text{Surface speed of delivery roller}}{\text{Surface speed of feed roller}}$$

Assume the delivery speed to be 1 rpm and draft change wheel = 1
Then,

$$\text{Draft} = \frac{\pi \times D \times 1}{\pi \times D \times \dfrac{30}{120} \times \dfrac{40}{\text{draft change wheel}}} = 0.1$$

But we know that

$$\text{draft} = \frac{\text{draft}}{\text{draft change wheel}} = 0.1$$

Hence draft constant = 0.1

Production calculations

Suppose the weight of one meter of the material is 55 g

$$\text{Hank} = \frac{8.33}{\text{Weight in grains/yard}}$$

$$= \frac{8.33}{55 \times 15.43/1.09} = 0.0107$$

Therefore delivery hank = 0.0107.

$$\text{Production in kg/shift} = \frac{\pi \times D \times N \times 60 \times 8 \times 80}{2.204 \times 36 \times 840 \times 100 \times \text{hank}}$$

$$= \frac{\pi \times 16.14 \times 40.92 \times 60 \times 8 \times 80}{36 \times 840 \times 2.204 \times 100 \times 0.0107}$$

$$= 1117.456 \text{ kg/shift.}$$

16.8 Comber

The main objectives of combing are as follows:

(i) Combing is done in order to increase the mean length of fibres.
(ii) Maximum possible elimination of short fibres can be well performed due to combing.

(iii) Lustre property of the fibre is slightly increased.
(iv) Parallelization of fibres is said to occur due to combing.

16.8.1 Methods of improving the noil percentage

I.   By changing the type of feed. Backward feed will obviously result in increasing no percentage because during backward feed of fibres combing is done 4 times where as it is 3 times in case of forward feed.

II. By changing the ratchet wheel number i.e. decreasing the ratchet wheel teeth amount of noil % can be increased.

III. By changing the distance between nipper to detaching or by changing the index. If the index is more, then the % of noils removed will also be more.

IV. By decreasing the distance between top comb and unicomb the percentage removal of noil is increased.

16.8.2 Technical specifications of a modern comber

No. of combing heads	8
Maximum mechanical speed	400 nips/min
Maximum attainable speed	350 nips/min
Delivery count	0.10 to 0.18 (Ne)
Lap dimension	267–305 mm width
Weight – max.	80 g/m
Actual feed rate per nip	3.76 to 5.91 mm
Type of feeding	Counter feed – standard
Forward feed	Optional
Maximum lap diameter	600 mm
Noil percentage	5–25%
Doubling	2 × 4 fold
Main motor	4 to 4.86 kW
Brush motor	1.5 kW
Can size diameter	400–450–500 mm
Height	1050 mm (42 in.)

$$\text{Combing efficiency} = \frac{50\% \text{ span length of comber sliver} - \text{lap}}{\text{Comber lap C}} \times \frac{100}{C}$$

where C is the noil %

Nep removal efficiency = Neps/g in – neps/g in × 100

$$\frac{\text{Comber lap} - \text{comber sliver}}{\text{Neps/g in comber lap}}$$

16.8.3 Gearing calculations of a comber

Speed calculations

(i) Speed of brush roller $= 2850 \times \dfrac{3.8}{9.5} \times \dfrac{16}{19} = 960$ rpm

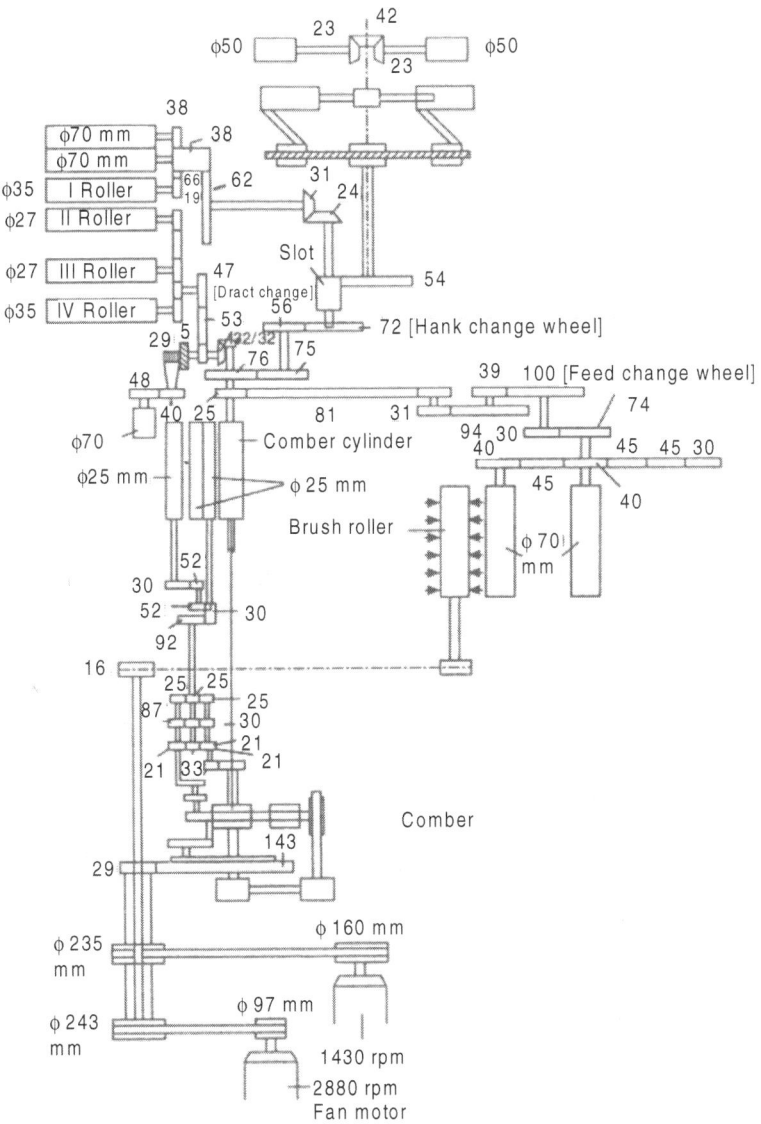

16.8 Gearing plan of a comber.

(ii) Speed of combing cylinder $= 1430 \times \dfrac{6.5}{5.5} \times \dfrac{29}{143} = 221.765$ rpm

(iii) Speed of table calender roller $= 1430 \times \dfrac{6.5}{8.5} \times \dfrac{29}{143} \times \dfrac{32}{32} \times \dfrac{5}{29} \times \dfrac{40}{48}$

$= 31.863$ rpm

(iv) Speed of third roller $= 1430 \times 143 = 84.9312$ rpm

16.8.4 Waste

The variation in comber waste between combers $= \pm\,0.5\%$
The variation in comber waste within a comber $= \pm 1.5\%$

16.9   Fly frame

The object of the fly or speed frame is the convert drawn or combed sliver into a thinner material called roving and also impart a little twist to it.

16.9.1 Technical specifications of modern speed frames

Production speed	maximum of 60 m/min
Processable fibres – cotton, blends	up to 60 mm fibre length
No. of spindles	36–48–60–72–84–96–108–120 and up to 144 in special cases
Mechanical speed of flyer	Maximum 1500 rpm
Roving package dimension	Diameter – 6 in. (152.4 mm)
Height	16 in. (406.4 mm)
Weight	2–3.2 kg
Feed count	Ne – 0.12 to 0.24
Delivery count	Ne – 0.4 to 3.5
Mechanical draft	4 to 20
Twist per inch	0.44 to 2.45
Twist per meter	17 to 96
Feed cans – diameter	16 in. – 18–20–24
Height	36–42–45
Drafting system	Pneumatic or SKF-PK-1500.

Optional – Incorporated doffing and bobbin transport system to ring frames (overhead)

16.9.2 Gearing calculations of a speed frame

Speed calculations

(i)   Speed of front roller $= 1430 \times \dfrac{165}{355} \times \dfrac{102}{74} \times \dfrac{32}{70} \times \dfrac{26}{67} = 162.52$ rpm

(ii) Speed of middle roller $=$ front roller speed $\times \dfrac{20}{130} \times \dfrac{48}{69} \times \dfrac{25}{23} =$
18 rpm

(iii) Speed   of   back   roller $=$ front   roller   speed $\times$
$\dfrac{20}{130} \times \dfrac{48}{69} \times \dfrac{45}{105} \times \dfrac{60}{35} \times \dfrac{22}{18} = 15.61$ rpm

(iv) Speed of spindle $= 1430 \times \dfrac{165}{355} \times \dfrac{102}{110} \times \dfrac{110}{82} \times \dfrac{38}{36} = 872.68$ rpm

(v)   Bobbin speed at initial position
Let
    a = differential box speed
    F = main shaft speed
    e = differential ratio
    $L = a + e\ (F - a)$

$a \;=\; 1430 \times \dfrac{165}{355} \times \dfrac{102}{74} \times \dfrac{32}{70} \times \dfrac{212}{108} \times \dfrac{27}{60} = 789.21$ rprn

$F \;=\; 1430 \times \dfrac{165}{355} = 664.647$

$e \;=\; \dfrac{59}{28} \times \dfrac{25}{62} = 0.85$

$L \;=\; 789.2150 + 0.85\ (664.67 - 789.2150)$
$\;=\; 620.44$ rpm

Therefore the bobbin speed at initial position

$= L \times \dfrac{82}{54} \times \dfrac{38}{36} \times 683.37 \times \dfrac{82}{54} \times \dfrac{38}{36} = 994.49$ rpm

At final position

$a = 1430 \times \dfrac{165}{355} \times \dfrac{102}{74} \times \dfrac{32}{70} \times \dfrac{108}{212} \times \dfrac{48}{50} = 204.819$ rpm

$L = 204.819 + 0.85\ [664.647 - 204.819] = 595.51$ rpm

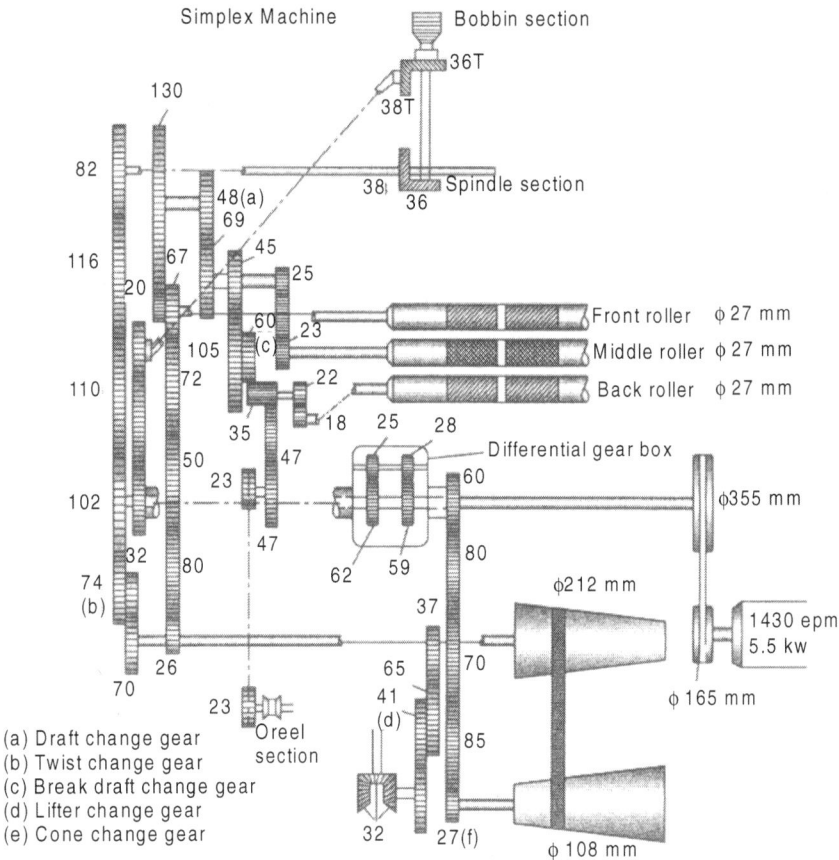

16.9 Gearing plan of a speed frame.

Therefore bobbin speed at final position $= 595.51 \times \dfrac{82}{54} \times \dfrac{38}{36}$

$$= 928.63 \text{ rpm.}$$

Draft calculations

(i) Front zone draft $= \dfrac{\text{Surface speed of front roller}}{\text{Surface speed of middle roller}}$

$$= \dfrac{\pi \times 27 \times 162.52}{\pi \times 27 \times 18.90}$$

$$= 8.59625$$

(ii) Back zone draft $= \dfrac{\text{Surface speed of middle roller}}{\text{Surface speed of back roller}}$

$$= \frac{\pi \times 27 \times 18.90}{\pi \times 27 \times 15.61}$$

$$= 1.210$$

(iii) Total draft = Front zone draft × Back zone draft

$$= 8.59625 \times 1.210 = 10.405$$

(iv) To find draft constant

$$\text{Draft} = \frac{\text{Surface speed of front roller}}{\text{Surface speed of back roller}}$$

Assume front roller speed = 1; Draft change pinion = 1
Then,

$$\text{Draft constant} = \frac{\pi \times 27 \times 1}{\pi \times 27 \times 1 \times \dfrac{20}{130} \times \dfrac{\text{DCW}}{69} \times \dfrac{45}{105} \times \dfrac{60}{35} \times \dfrac{22}{18}}$$

$$= 499.465$$

(v) Draft checking:

$$\text{Draft} = \frac{\text{Draft constant}}{\text{Draft change wheel}} = \frac{499.465}{48} = 10.405$$

Twist calculations

$$\text{Twist per inch} = \frac{\text{Spindle speed}}{\text{Front roller delivery in inches/mm}}$$

$$= \frac{872.86}{\pi \times \dfrac{27}{25.4} \times 162.52}$$

To find twist constant:
Assume spindle speed = 1 rpm
Twist change wheel 1 teeth
Then,

$$\text{Twist constant} = \frac{1}{1 \times \dfrac{36}{38} \times \dfrac{82}{\text{TCW}} \times \dfrac{32}{70} \times \dfrac{26}{67} \times \pi \times \dfrac{27}{25.4}}$$

$$= 0.0217$$

Twist constant = Twist change wheel × twist constant

$$= 74 \times 0.0217$$

$$= 1.6079$$

Production calculations

Assume that hank delivered = 1.3
Efficiency = 8%
Then,

$$\text{Production/shift/spindle} = \frac{\pi \times 1.0629 \times 162{,}52 \times 80 \times 8 \times 60}{840 \times 36 \times 2.204 \times 1.3 \times 100}$$

$$= 2.405 \text{ kg/shift/spindle}$$

$$\text{Total production} = 2.405 \times 36$$

$$= 86.59 \text{ kg/shift/frame}$$

16.10   Ring frame

The objectives of the ring frame are three fold:

(a) To draft the input material to the linear density required in the final yarn,
(b) To insert the required amount of twist, and
(c) To wind the yarn onto a package which is suitable for handling, storage and transport and is capable of being unwound at high speed in subsequent processing.

16.10.1 Technical specifications of modern ring frame

Roller stand angle	– 45°
Distance between center of front	– 66.5 mm
roller nip to lappet hook centre	
Gauge 70 mm	– Tube length: 180; 190; 200 mm
70– 75 mm	– Tube length: 201; 220; 230 mm
75 mm	– Tube length: 240; 250; 260 mm.
Mechanical speed	– Maximum 25,000 rpm
Count range	– Ne 5.5 to 120s
Twist per inch	– 4 to 55
Draft system	– Double apron 3-roller to arm
No. of spindles	– from 96 to 1008
Motors main	– Minimum of 15 kW to Maximum of 37 kW 45kW with inverter
Ring lowering	– 0.35 kW
Suction	– 0.35 kW
Pantograph drive	– 2.2 kW

Geared motors:

Horizontal transport belt	– 2 × 0.12 kW
Inclined transport belt	– 2 × 0.185 kW

Tube feeding	– 2.5 minutes
Doffing time	– 2.5 minutes
Bobbin discharge speed	– 60 (30 + 30) minutes

Compressed air consumption:

For tube fitting on the band	– 1.35 litre per minute
For bobbin doffing operation	– 0.40 litre per spindle
Ring diameter	– 38–40–42–45–48 mm
Cradle length	– 36–43–59 mm. Pneumatic 40–51–60 mm for SKF drafting

16.10.2 Miscellaneous formulae

$$\text{Actual Count} = \frac{7000}{\text{Weight in grains of yard (1 lb = 7000 Grains)}}$$

$$\text{Corrected count} = \frac{\text{Actual count (1 + present region)}}{(1 + \text{standard regain})}$$

$$\text{Actual Count} = \frac{7000}{\text{Weight in grains} \times 7} \quad (7 \text{ leas} = 1 \text{ hank})$$

Yarn realization

$$\text{Carded yarn} = 93 - 1.2 \, t$$

where t = trash in cotton

$$\text{Combed yarn} = \frac{(96 - 1.2 \, t)(1 - W) - 3}{100}$$

$$\text{Process proficiency} = \frac{\text{Actual CSP}}{\text{Expected CSP}} \times 100$$

$$\text{Standard hank} = \frac{\text{Actual spindle speed} \times (1 - \text{doff loss})}{\text{TPI} \times 63}$$

$$\text{Total draft} = \frac{\text{Count spun}}{\text{Hank fed}}$$

$$\% \text{ Contraction} = \frac{(\text{Mechanical draft} - \text{Actual draft})}{\text{Mechanical draft}} \times 100$$

(or)

$$\frac{(\text{Count of roller nip} - \text{Count on the bobbin})}{\text{Count at the roller nip}} \times 100$$

$$\text{Actual draft} = \frac{\text{Count spun}}{\text{Hank of roving}}$$

$$\text{Mechanical draft} = \frac{\text{Count at roller nip}}{\text{Hank of roving}}$$

$$\text{Change pinion} = \frac{\text{Draft constant}}{\text{Mechanical draft}}$$

16.10.3 Waste

The recommended norms for various types of waste in ring spinning are given in table below:

Type of waste	Good	Average	Poor
Yarn waste	0.3	0.5	0.8
Sweep waste	1.3	1.7	2.0
Invisible loss	0.5	1.2	2.0

The recommended norms of usable waste (%) for good working are given in table below:

Type of waste	Lap fed system	Chute feed system
Lap bits and card web	0.7	0.2
Sliver in drawing and fly frames	0.5	0.5
Waste at comber preparatory and comber	1	1
Roving ends	0.3	0.3
Pneumafil and bonda waste	5	4.5

16.10.4 Gearing calculations of a ring frame

Speed calculations

(i) speed of front roller
$$= 1440 \times \frac{223}{404} \times \frac{26}{115} \times \frac{83}{82} \times \frac{72}{65} \times \frac{30}{47} \times \frac{103}{102} \times \frac{26}{29}$$
$$= 124.37 \text{ rpm}$$

(ii) speed of middle roller
$$= 1440 \times \frac{223}{404} \times \frac{26}{115} \times \frac{83}{82} \times \frac{73}{63} \times \frac{30}{44} \times \frac{103}{102} \times$$
$$\frac{70}{103} \times \frac{70}{112} \times \frac{52}{49}$$
$$= 10.09 \text{ rpm}$$

(iii) speed of back roller $= 1440 \times \dfrac{223}{404} \times \dfrac{26}{115} \times \dfrac{83}{82} \times \dfrac{75}{65} \times \dfrac{30}{44} \times$

$$\dfrac{103}{102} \times \dfrac{70}{103} \times \dfrac{20}{112} \times \dfrac{21}{47}$$

$= 7.522$ rpm

(iv) speed of cam $= 1440 \times \dfrac{223}{404} \times \dfrac{26}{115} \times \dfrac{83}{82} \times \dfrac{72}{65} \times \dfrac{30}{44} \times$

$$\dfrac{103}{102} \times \dfrac{70}{64} \times \dfrac{20}{50} \times \dfrac{21}{121}$$

$= 10.667$ rpm

Draft calculations

(i) Front zone draft $= \dfrac{\text{Surface speed of front roller}}{\text{Surface speed of middle roller}}$

$$= \dfrac{\pi \times 27 \times 124.37}{\pi \times 27 \times 10.667} = 11.3166$$

(ii) Back zone draft $= \dfrac{\text{Surface speed of middle roller}}{\text{Surface speed of back roller}}$

$$= \dfrac{\pi \times 27 \times 10.99}{\pi \times 27 \times 7.522} = 1.467$$

(iii) To find draft constant

Draft constant $= \dfrac{\text{Surface speed of front roller}}{\text{Surface speed of back roller}}$

Assume front roller speed = 1 rpm
Draft change wheel = 1

Draft constant $= \dfrac{\pi \times 27 \times 1}{\pi \times 27 \times 1 \times \dfrac{29}{26} \times \dfrac{\text{DCW}}{103} \times \dfrac{20}{112} \times \dfrac{21}{47}}$

(iv) Total draft $= \dfrac{\text{Draft constant}}{\text{DCW}} = \dfrac{1218.30}{70} = 17.4$

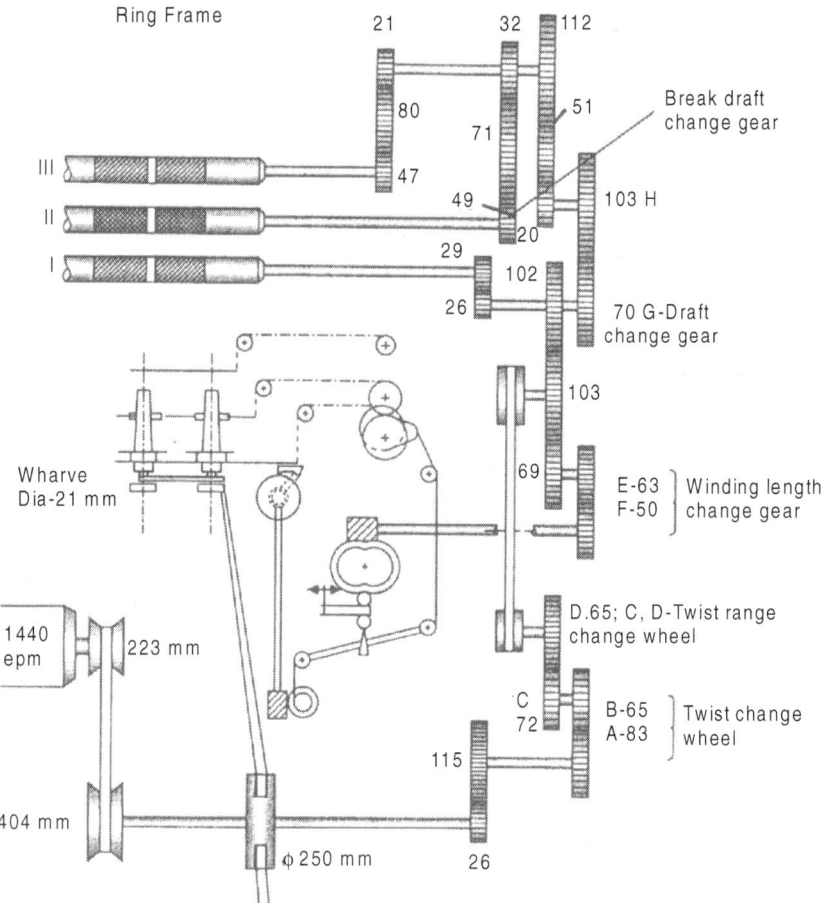

Ring Frame

16.10 Gearing plan of a ring frame.

Twist calculations

$$\text{Twist per inch} = \frac{\text{Spindle speed}}{\text{Front roller delivery}}$$

$$= \frac{9462.51}{\pi \times \dfrac{27}{25.4} \times 124.37} = 22.78$$

$$\text{Twist constant} = \frac{\text{Spindle speed}}{\text{Delivery speed in inch/min}}$$

$$= \frac{9462.51}{\pi \times \dfrac{27}{25.4} \times 124.37} = 22.78$$

$$\text{Twist constant} = 22.78 \times 83 = 1890.74$$

Assume twist change wheel = 1

Then,

$$\text{Twist constant} = \cfrac{9462.51}{\pi \times \dfrac{27}{25.4} \times 1440 \times \dfrac{223}{104} \times \dfrac{26}{115} \times \dfrac{1}{82} \times \dfrac{72}{65} \times \dfrac{30}{44} \times \dfrac{103}{102} \times \dfrac{26}{29}}$$

$$= 1890.94$$

Production calculation

$$\text{Production/shift/machine} = \frac{\pi DN \times 8 \times 60 \times 95}{TPI \times 36 \times 840 \times 2.204 \times 100 \times \text{count}}$$

$$= \frac{\pi \times 1.062 \times 124.37 \times 8 \times 60 \times 95}{36 \times 840 \times 20 \times 2.204 \times 100}$$

$$= 0.1419 \times 64 = 9.0847 \text{ kg/shift/machine}$$

Hours of operative per 100 Kg of production (HOK) $= \dfrac{N_1 \times N_2}{P} \times 100$

Where N_1 = Total no. of workers in a given section

N_2 = No. of working hours per month

16.11 Basic requirements for a standard spinning mill

As per SITRA norms a standard spinning mill should meet the following requirements:

(a) Production/spindle/8 hours adjusted to 40s = 88.8 g

(b) HOK – adjusted to 40s; operative

Hours to produce 100 kg of yarn:
Up to spinning = 24.0
Post spinning = 9.7

(c) Spindle shift per annum = 1000

(d) UKG – units of power per kg of yarn
Adjusted to 40s = 3.56

Countwise production requirements for a standard spinning mill

Count	Production/spindle in g/8 h
20s	212.8
30s	129.8
40s	88.8
60s	52.1
80s	34.9
100s	24.1

16.12 Conversion cost per spindle shift

$$\text{For cone yarn} = \frac{16.3}{\sqrt{\text{Count}}}$$

$$\text{For hank yarn} = \frac{18}{\sqrt{\text{Count}}}$$

16.13 General causes for end breakages in ring spinning

(i) Exhausted bobbin

(ii) Breakage of roving

(iii) Due to undrafted material

(iv) Thick places in roving

(v) Weak places in-roving

(vi) Lapping on top or bottom roller

(vii) Traveler missing/fly out

(viii) Incorrect setting of traveler clearer

(ix) Incorrect traveler

(x) Damage of rings

(xi) Lashing separators not effective

(xii) Spindle defect (vibration/wobbling/incorrect center)

(xiii) Double roving

(xiv) Damaged or missing apron

(xv) Top roller cot problem

(xvi) Use of incorrect spacers

(xvii) Misalignment of spindle tapes

(xviii) Improper thread guide setting

(xix) Improper humidity condition

(xx) Improper traversing mechanism

(xxi) Excessive soft waste in mixing

16.14 Reasons for high end breakage rates

(i) Tenters not trained well for piecing or unskilled labourers.
(ii) Inferior quality of roving material
(iii) Improper position of creel which exerts more tension to the roving
(iv) Improper selection of ring and traveler, and
(v) High humidity in the department.

Calculations in yarn testing

17.1 Introduction

This chapter deals with various formulae relating to yarn testing. Count and twist are the two important parameters governing the properties of a yarn. Both are of prime importance in the design of textile structures and, to a large extent, they govern the appearance and behaviour of the various types of yarns and fabrics. The count of a yarn is a numerical expression which defines its fineness. In other words it is the mass per unit length or length per unit mass. When the geometry of the fabric is considered, the diameter of the yarn has to be taken into account. For this the yarn is assumed to be circular in cross section and the diameter is calculated from the count. A number of simple formulae are available for estimating the yarn diameter. Twist is the measure of the spiral turns given to a yarn in order to hold the constituent fibres or threads together. Twist is necessary to give a yarn coherence and strength.

17.2 Miscellaneous formulae

Tunis per inch = twist factor × √count
Tex twist factor = turns per metre √n
In case of a tensile testing instrument, the breaking load is given by

$$F_T = F_{10} (1.1 - 0.1 \log T)$$

i.e. $\qquad \dfrac{F_T - F_{10}}{10} = 0.1 \log (10/T)$

where, F_T = breaking load
T = time to break in seconds
F_{10} = breaking load for a time to break of 10 seconds.

In a single thread strength tester working on CRT principle

$$\text{Tenacity in grams per tex} = \frac{\text{Mean breaking load in kg}}{\text{Tex}} \times 1000$$

$$\text{Elongation \%} = \frac{\text{Elongation scale reading}}{\text{Gauge length}} \times 100$$

In a ballistic tester,

$$\text{Work of rupture} = W\,(h_1 - h_2) \text{ in. lb}$$

Where, W = weight of the pendulum
h_1 = height of the pendulum above its lowest point
h_2 = height of pendulum after yarn break

Basic irregularity or limit irregularity is given by

$$Vr = \frac{100^2}{N} + \frac{Vm^2}{N}$$

Where, Vr = CV% of weight per unit length of the strand
N = average number of fibres in the cross section of the strand
Vm = CV% of the fibre weight per unit length

Index of Irregularity,

$$I = \frac{Va}{Vr}$$

Where, I = index of irregularity
Va = the actual irregularity measured
Vr = the calculated limit irregularity

For cotton,

$$Vr^2 = \frac{(106)^2}{N}$$

The number of fibres in the cross section of a strand is given by

$$N = \frac{15{,}000}{\text{Count} \times \text{fineness in micrograms per inch}}$$

Where N is no. of fibres in the cross section of a strand
For the Tex count,

$$N = \frac{1000\,T}{H}$$

Where, T = count in tex
H = the fibre weight in milligrams per kilometer

Addition of irregularities

$$V^2 = (V_1)^2 + (V_2)^2$$

Where, V – total irregularity;
V_1 – original irregularity;
V_2 – added irregularity.

$$CV \text{ of a doubled strand} = \frac{CV \text{ of individual strand}}{\sqrt{n}}$$

n – number of strands

In the analysis of periodic variation, the following formula can be used to find out the Wave length of the periodic variation introduced by the previous machine. Thus, wave length of variation introduced by previous machine

$$= \frac{\text{Wavelength of variation in present machine}}{\text{Draft in present machine}}$$

Thus periodic variation introduced by simplex is given by,

$$\text{Wave length of variation introduced by simplex} = \frac{\text{Wavelength of variation in yarn}}{\text{draft in ring frame}}$$

Similarly, the periodic variation introduced by draw frame is,

Wave length of variation introduced by draw frame

$$= \frac{\text{Wavelength of variation in yarn}}{\text{draft in ring frame}}$$

Formula relating wave length, draft and the diameter of the roller producing the fault is given as,

Wave length of fault in inches

$$= \pi D \times \text{draft between the faulty roller and the front roller}$$

where, D – diameter of the faulty roller in inches

An alternative method of locating a faulty roller is based upon the formula,

Wavelength of variation in inches =

$$\frac{\text{delivery speed of front roller in inches/mm}}{\text{faulty roller speed in rpm}}$$

Conversion of Uster to Fielden:

Slivers ULI = O.42F + 2.85 = 1.17F (sample length 250 ft and speed
               of Testing 20 YPM)
Roving ULI = 0.57F + 2.8 = 0.86 (sample length 500 ft and testing speed
               100 YPM)
yarns ULI   = 1.04F + 1.1 = 1.2F (sample length 1000 ft and testing
               speed 20 YPM)
ULI   = Uster linear integrator
F   = Fielden walker

17.3   Basic definitions

Yarn number

The yarn number is a measure of the linear density of the yarn. It can be
expressed in various ways in terms of direct or indirect or indirect system.
In the direct system, the yarn number is the weight per unit length of a
yarn and in the indirect system it is the length per unit weight of yarn.

Count

It is a measure of the linear density of yarn considering length of a yarn in
one lb. The unit of length is 840 yards.

For example, yarn of 840 yards weighing one lb is count and yarn of
8400 yards weighing one lb is 10s count.

Tex

It is a measure of the linear density of yarn considering weight in grams in
a thousand meters length of the yarn. The unit of weight is one gram.

For example, yarn weighing one gram in a length of thousand meters is
one tex and yarn weighing 5 g in a length of thousand meters is 5 tex.

Denier

It is a measure of the linear density of yarn considering weight in grams in
a length of 9000 meters. The unit of weight is one gram.

For example, yarn weighing one gram in a length of 9000 meters one
denier and yarn weighing 10 grams in a length of 9000 meters is 10 denier.

17.4   Yarn numbering systems

There are two systems of yarn numbering, namely,

1. Indirect or fixed weight system, and
2. Direct or fixed length system

In the indirect system the count of yarn expresses the number of length units in one weight unit. Hence higher the count finer is the yarn. This system is generally used for cotton, worsted, woolen, linen (wet spun), etc.

The direct system expresses the yarn count as the number of weight Units in one length unit. Hence higher the count, coarser is the yarn. This system is suitable for thrown silk, artificial silk, jute, etc.

Different units of length and weight are used for different textile yarns. Sometimes even for the same kind of yarn, different Units of length and weight are used in different

17.5   Universal system of yarn numbering

It was thought of to evolve a system of yarn numbering that would be suitable for all types of yarns. The tex system has got approval of the international standardization organization (ISO) for adoption as the universal count system.

In yarn count and related calculations, three factors are to be considered, namely, length, weight and count of yarns. The formula for count has been deduced, in each case from the basic principle. From this, all other related calculations have been formulated by using simple mathematical rules.

17.6   Practical significance of yarn count

The yarn count is helpful in the quantitative estimation of yarns required or contained in different types of textile materials, like folded yarns, warps and cloths as well as where the single yarns, folded yarns and warps are in the form of cones, cheeses, bundles, balls, chains and beams, as applicable. Another important use is in the calculations involved in certain aspects of production of the preparatory department and weaving shed. It is also required for calculations pertaining to yarn diameter, cloth setting, cloth cover, etc.

17.7   Indirect system of yarn numbering

As already defined earlier, in this system the count is based on the number of length units in one weight unit. This system is suitable for most of the important types of yarns.

For example, if 30 length units weigh one weight unit, its count is 30s, and if 60 length units weigh one weight unit then its count is 60s.

Thus it can be seen that finer or less bulky the yarn, higher is its count number or **iii** other words, the size or bulkiness of the yarn is inversely proportional to the count number and hence the name indirect system.

The disadvantages of this system are:

(a) The count number does not directly express the size of the yarn.
(b) In count calculations of folded yarns, it is more difficult to calculate the resultant count, especially of different counts.

The advantages of the system are the following:

(a) The weight of the fabrics can be calculated with comparative ease.
(b) The twist of yarn, setting of cloth, etc., can be more easily calculated as they are more directly proportional to the square root of the counts of the yarn.

The table below shows the different length and weight units used in important indirect systems.

Name of the system	Unit of weight	Unit of length
English cotton	1 lb	Hank of 840 yards
French cotton	1/2 kg	Hank of 1000 meters
Metric cotton	1 kg	Hank of 1000 meters
Spun silk	1 lb	Hank of 840 yards
Worsted	1 lb	Hank of 560 yards
Linen (wet spun)	1 lb	Lea of 300 yards
Hemp (fine)	1 lb	Lea of 300 yards
Woolen – Yorkshire skein	6 lbs	Skein of 1,536 yards
Woolen – American cut	1 lb	Cut of 300 yards
Asbestos –British	1 lb	Hank of 50 yards
Asbestos –American	1 lb	Cut of 100 yards
Fibre glass	1 lb	Cut of 100 yards

According to the indirect system,

$$\text{Count} = \text{Number of length units per weight unit}$$

Or

$$\text{Count} = \frac{\text{Length in appropriate length unit}}{\text{Weight in corresponding weight unit}} \qquad \text{... (a)}$$

By cross multiplication, we have the following:

$$\text{Weight} = \frac{\text{Length in appropriate length unit}}{\text{Count}} \qquad \text{... (b)}$$

$$\text{Length} = \text{Count} \times \text{weight, in appropriate length unit} \qquad \text{... (c)}$$

From the above it can be seen that there are three factors, viz., count, length and weight, of which two must be give to find out the third one.

17.8   Direct system of yarn numbering

In this system the count is based on the number of weight units in a length unit. Thus the size or bulkiness of the yarn is directly proportional to the count number. A coarser yarn will have a higher count number while a finer yarn will have a lower count number. The system is used for very fine and very coarse yarns.

The table below shows the length and weight units used in important direct systems.

Name of the system	Unit of weight	Unit of length
Rayon, silk, thrown – denier English	1 denier	Hank of 520 yards
Rayon, silk, thrown – denier metric	0.05 g	450 meters
Nylon, vinyon etc. – denier metric	0.05 g	450 meters
Dram	1 dram	Hank of 1000 yards
Flax (Dry spun)	1 lb.	Spyndle of 14,400 yards
Jute	1 lb.	Spyndle of 14,400 yards
Hemp	1 lb.	Spyndle of 14,400 yards
Woolen USA grain	1 grain	20 yards
Woolen Aberdeen	1 lb	14,400 yards

According to the direct system of yarn numbering,

$$\text{Count} = \frac{\text{Weight in appropriate unit}}{\text{Length in appropriate unit}} \quad \dots \text{(a)}$$

Cross multiplication of the above gives

$$\text{Length in appropriate unit} = \frac{\text{Weight in appropriate unit}}{\text{Count}} \quad \dots \text{(b)}$$

Weight in appropriate unit = Length in appropriate unit × count   ... (c)

17.9   Conversion from one system to another in the in direct system

On certain occasions it may be necessary to know the equivalent count of one system in another. In such a situation the following method may be used to ascertain the count in the required system.

Method

Convert the length and weight units of the known system into the length and weight units respectively of the required system. Then divide the converted length unit by the converted weight unit. The result multiplied by the count of the known system will give the count in the required system.

Count in required system

$$= \frac{\text{Count in known system/Count in required system}}{\text{Weight unit in known system/Weight unit in required system}}$$

For example, convert 40s English cotton count to worsted count.

$$\frac{\text{Length unit in English cotton}}{\text{Length unit in worsted}} = \frac{840}{560} \text{ yards}$$

$$\frac{\text{Weight unit in English cotton}}{\text{Weight unit in worsted 1}} = 1 \text{ lb}$$

$$\text{Therefore, required count} = 40 \times \frac{840/560}{1/1} = 60$$

17.10 Conversion factors for the indirect system

Required count	Cotton	Spun silk	Metric	Linen	Wool	Worsted (Yorkshire)
Cotton	1	1	1.693	2.8	3.281	1.5
Spun silk	1	1	1.693	2.8	3.281	1.5
Metric system	0.591	0.591	1	1.654	1.938	0.886
Linen and hemp (fine)	0.357	0.357	0.605	1	1.172	0.536
Woolen (Yorkshire)	0.305	0.305	0.516	0.853	1	0.457
Worsted	0.667	0.667	1.129	1.867	2.188	1

The various conversion factors to be used to convert yarn count from one system to another, both in the indirect system, arc shown in table below:

To convert counts in any system in the 1st vertical column to any system in the 1st horizontal column of the table, multiply the corresponding conversion factor by the known count. The product is the count in the required system.

$$\text{Required count} = \text{Known count} \times \text{Conversion factor}$$

For example, convert 30s cotton into spun silk count.

$$\text{Spun silk count} = \text{cotton count} \times \text{conversion factor}$$
$$= 30 \times 1 = 30$$

17.11 Conversion factors for the direct system

Whenever the equivalent count of one system is required in another system, the following method is adopted for converting the known count into the required one.

Method

Convert the length and weight units of the known system into the length and weight units respectively of the required system. Then divide the converted weight unit by the converted length unit. The result multiplied by the count of the known system will give the count in the required system.

Count in required system

$$= \frac{\text{Weight unit in known system/Weight unit in required system}}{\text{Length unit in known system/Length unit in required system}}$$

For example, convert 30s Rayon (denier metric) into flax (dry spun) count.

$$\text{Required count} = \frac{0.05/453.6 \times 30}{450/14400 \times 09144} = 0.0965$$

Conversion factors for determining equivalent count of one system in another in the direct system are given in table below:

Required count	Silk (metric denier)	Dram	Jute spindle	Flax spyndle	Woolen grain	Woolen (Aberdeen)
Silk denier (metric)	1	0.05734	0.003225	0.003225	0.03136	0.003225
Silk (dram)	17.44	1	0.05625	0.05625	0.5469	0.05625
Jute (spyndle)	310	17.78	1	1	9.722	1
Flax (spyndle)	310	17.78	1	1	9.722	1
Woolen(grain)	31.89	1.829	0.1029	0.1029	1	0.1029
Woolen Aberdeen	310	17.78	1	1	9.722	1

To determine the equivalent count of any one system in the 1st vertical column in any system in the top horizontal column, multiply the known count by the conversion factor.

The product is the equivalent count in the required system.

Required count = known count × conversion factor.

For example, what is the equivalent count of 20s woolen (grain) in dram. Conversion factor is 1.829

17.12   Conversion from the indirect system to the direct system and vice versa

Therefore, equivalent count = 20 × 1.829 = 36.58

Methods used in converting counts from direct system to indirect system and vice versa are given below:

From the indirect to the direct system

In this method the length and weight units of the known indirect system are converted into the length and weight units, respectively, of the required direct system. Then the converted weight unit is divided by the converted length unit. The result divided by the known count of the indirect system will give the required count in the direct system.

Count in required system

$$= \frac{\text{Weight unit in known system/weight unit in required system}}{\text{Length unit in known system/length unit in required system}} \times$$

$$\frac{1}{\text{count in known system}}$$

For example, convert 60s cotton count into denier (metric) system.

$$\text{Count in required system} = \frac{453.6/0.05}{840/450 \times 0.9144} \times \frac{1}{60} = 88.58$$

17.12.1 From the direct to the indirect system

In this system the length and weight units of the known direct system is converted into the length and weight units, respectively, of the required indirect system. The converted length unit is divided by the converted weight unit. The result divided by the known count of the direct system will give the required count in the indirect system. Count in required system

$$= \frac{\text{Length unit in known system/length unit in required system}}{\text{Weight unit in known system/weight unit in required system}} \times$$

$$\frac{1}{\text{count in known system}}$$

For example, convert 88 denier silk to cotton count.

$$\text{Count of cotton} = \frac{\text{Length in denier system/length in cotton system}}{\text{Weight in denier system/weight in cotton system}} \times$$

$$\frac{1}{\text{count in denier system}}$$

$$= \frac{9000/840 \times 0.9144}{1/453.6} \times \frac{1}{88}$$

17.12.2 Count calculations in tex system

Count in Tex system

$$= \text{weight in grams/length in km} = \frac{\text{Weight in grams}}{\text{length in meters}/1000}$$

$$= \frac{1000 \times \text{weight in grams}}{\text{length in meters}}$$

$$\text{Length} = \frac{\text{Weight in grams}}{\text{count}}$$

or $\qquad \text{Weight in grams} \times \dfrac{1000}{\text{count}}$

$\qquad\qquad \text{Weight} = \text{count} \times \text{length in km}$

Or $\qquad 1000 \times \text{count} \times \text{length in meters}$

17.12.3 Conversion from indirect system to universal tex system

Cotton (English)	590.5
Spun silk and spun rayon staples	590.5
Metric	1000.0
Linen (wet spun)	1653.0
Worsted	885.8
Woolen Yorkshire skein	1938.0

$$\text{Required count} = \frac{\text{Conversion factor}}{\text{Known count}}$$

17.12.4 Conversion from direct systems to universal tex system

Silk, denier (metric)	0.111
Silk dram	1.938
Jute, spyndle	34.450
Flax spyndle	34.450

| Woolen, Aberdeen | 34.450 |
| American grain | 3.543 |

Required count = Known count × Conversion factor

17.13   Calculations pertaining to ply or folded yarns in the indirect system

The counts of folded yarns in cotton, linen and worsted are expressed in terms of the counts of the single threads of which the folded yarn is composed. Thus 2/60s implies a ply yarn made by twisting together two 60s single threads and is equivalent to 30s. There will be some contraction in length when two threads are twisted together, So that the actual count of single threads required to produce a two ply yarn of 30s count will be slightly finer than 60s, This, however, is not significant in normal cases. Where it is significant, the contraction due to twisting should be taken into account.

In case of fancy yarns, the component single threads of which the folded yarn is composed arc of different counts. In such cases the count of the ply yarn is expressed in terms of the component single threads.

Thus 24/40 expresses a folded yarn made by twisting together two single threads, one of 24s and other of 40s and is equivalent to 15s and not the arithmetic mean of 24 and 40. This is illustrated in an example below.

In the case of spun silk the count of the folded yarn is expressed in a different manner:

Thus 100/2 represents a yarn made by twisting together two single threads and the count of the resultant yarn is l00s.

In the case of yarns, whose counts are expressed in denier, no mention is generally made of the folding number.

Method of finding the count of the resultant ply yarn when the counts of the component threads of which it is made are known.

Assume a definite length unit, say, one hank. Then the weight of such length unit of each component threads is found out. The sum of these weights will be the weights will be the weight of the assumed length unit of the resultant folded yarn. The length and the weight being known, the count of the folded yarn can be easily ascertained by using any suitable formula explained below.

In the case of ordinary ply yarn where the contraction due to twisting is negligible, the following formulae may be used.

Resultant count (indirect system)

$$= \frac{1}{\text{Weight in lb. of 1 length unit of folded yarn}}$$

$$= \frac{1}{\text{Weight in lb. of 1 length unit A + weight in lb of}}$$
$$\text{1 length unit of B + \&\&}$$

Where A, B,…are component threads.

For example, calculate the count of the two fold cotton yarn composed of 20s and 40s singles.

$$1 \text{ hank of 20s} = \frac{1}{20}lb$$

$$1 \text{ hank of 40s} = \frac{1}{40}$$

$$1 \text{ hank of folded yarn} = \frac{1}{20} + \frac{1}{40} = \frac{3}{40}lb$$

Therefore, resultant count of folded yarn

$$= \frac{1}{\text{Weight in lb. of 1 hank of folded yarn}}$$

$$= \frac{1}{3/40} = 13s = 13s \text{ cotton (approx)}$$

For example, calculate the count of the three fold cotton yarn composed of 20s, 16s and 10s singles.

$$1 \text{ hank of 20s} = \frac{1}{20} lb$$

$$1 \text{ hank of 16s} = \frac{1}{16} lb$$

$$1 \text{ hank of 10s} = \frac{1}{10} lb$$

Therefore, 1 hank of folded yarn $= \frac{1}{20} + \frac{1}{16} + \frac{1}{10} = \frac{13}{80}$ lb

Therefore, resultant count of folded yarn

$$= \frac{1}{\text{Weight in lb. of 1 hank of folded yarn}} = \frac{1}{13/40} = 3s \text{ cotton}$$

17.13.1 Methods of finding the count

In this section, we discuss the method of finding the count of one of the unknown component threads, when the counts of the other components and the resultant folded yarn are known.

In this method a definite length (e.g., one hank) may be assumed. Then the weight of such a length unit of the unknown component can be found out by subtracting the sum

of weights of the assumed length unit of the resultant folded yarn. Then the length and weight of the unknown component thread being known its count can be easily ascertained. Count of unknown component thread (indirect system)

$$= \frac{1}{\text{Weight in lb of length unit of the unknown component thread}}$$

$$= \frac{1}{\text{Weight in lb of 1 length unit of folded yarn} - \text{Weight in lb. of}}$$
$$\text{1 length unit of component known threads}$$

For example, one thread of an unknown count, when plied together with another of 64s cotton, gives a two-fold yarn of 28s cotton. Calculate the count of the unknown thread.

1 hank of folded yarn = $\dfrac{1}{28}$ lb

1 hank of known single thread = $\dfrac{1}{64}$ lb

Therefore, 1 hank of unknown single thread = $\dfrac{1}{28} - \dfrac{1}{64} = \dfrac{23}{448}$

$$= 20\text{s } (approx)$$

For example, a three-fold cotton yarn composed of 6s, 20s and a thread of unknown count was found to be of 5s cotton. Calculate the count of the unknown thread.

1 hank of 3-fold yarn = $\dfrac{1}{5}$ lb

1 hank of 6s yarn   = $\dfrac{1}{6}$

1 hank of 20s yarn   = $\dfrac{1}{20}$

Therefore, 1 hank of unknown thread = $\dfrac{1}{5} - \left\{ \dfrac{1}{6} + \dfrac{1}{20} \right\} = \dfrac{1}{60}$

Therefore, count of unknown component yarn
'Weight in lb. of 1 hank of the unknown component thread

17.13.2 Calculations pertaining to ply or folded yarns in the direct system

The calculations in this system are relatively simpler compared to the previous one. This is due to the fact that the count number in these systems expresses the weight expresses the weight of one length unit of the yarn or thread. The weight of one unit length of folded yarn is equal to the sum of the weights of its component threads in one length unit. Therefore, the count of the folded yarn will also be the sum of the count numbers of its component threads.

Thus the resultant count of a two-fold yarn, composed of two threads of 60 denier and the other of 80 denier, will be the sum of the count number of these two component threads, i.e. $60 + 80 = 140$ denier.

As already mentioned previously, there will some contraction in length when the component threads are twisted together. The resultant count number in the above two examples will be higher than the sum of the count numbers of the component threads. The reverse also holds good. Thus if a two-fold yarn of 50 tex is composed of two single threads of similar counts, then the count number of each component will be slightly less or finer than 50/2 or 25 tex. This is not significant for ordinary cases. However, in the case of hard twisted yarns, special or fancy yarns, e.g. crepe yarns, corkscrew yarns loop yarns, etc., this factor is very significant and must be taken into account. The count of the folded yarn in such cases will be much higher than the sum of the count numbers of the component threads.

17.13.3 Calculation of the resultant count of the folded yarn

In this section, we discuss the calculation of the resultant count of the folded yarn the when the counts of the components are known.

As explained above, the count of the folded yarn will be the sum of the count numbers of the component threads.

Resultant count of the folded yarn = Weight of 1 length unit of folded
yarn in appropriate unit.

Total weight in appropriate unit of 1 length unit of component threads, A, B, C

Where, A,B C are component threads ... (a)
Count of A plus count of B plus count of C ... (b)

For example, calculate the count of a 2-fold polyester yarn, if the counts of its two component threads are 32 tex and 44 tex.

The resultant count of 2-fold yarn = 32 + 44 = 76 tex.

17.13.4 Calculation of the count

In this section, we discuss the calculation of the count of one of the unknown component thread when the counts of other components and the resultant folded yarn are known.

Since the count expresses the weight in appropriate unit of one length unit of yarn or thread, the count of the unknown component will be equal to the count of the folded yarn minus the sum of the count numbers of the known component threads.

Count of the unknown component thread
= Weight in appropriate unit of one length unit of the folded yarn – sum of the weights in appropriate unit of 1 length unit each of A, B

Where A, B . . . are the known component threads
= Count of the folded yarn – (count of A + count of B)

For example, calculate the count of the unknown component thread of a 2-fold yarn if the count of the folded yarn is 20 denier and that of the known component is 10 denier.

Applying the above formula, we have

Count of the component thread = Count of folded yarn count of known component, thread = 20 – 10 = 10 denier.

17.14 Conversion of worsted and rayon yarn count into cotton counts

$$\text{Worsted} = \frac{\text{Unit of length in worsted}}{\text{Unit of length in cotton count}} \times \text{Worsted count}$$

i.e. $$\frac{560 \times X}{840} \qquad \ldots \text{(a)}$$

$$\text{Denier} = \frac{9000 \times 1000 \times 1.1}{2.2 \times \text{denier} \times 840} \qquad \ldots \text{(b)}$$

Therefore count is $$\frac{1}{\text{cotton yarn count}} + \frac{1}{a} + \frac{1}{b}$$

Yards/lb for 1s denier = Yards/hank in silk × denier per ounce ×
ounces per lb

$$\text{Diameter of fiber in microns} = \frac{2000\sqrt{A}}{\sqrt{B\pi 9000}}$$

where   A – fibre denier
B – fibre specific gravity

or $\dfrac{20\sqrt{\text{denier}}}{\sqrt{3.14 \times \text{specific gravity}}}$

Number of fibres per cross section

$$= \frac{53.15 \times \text{sum of (\% of each denier) denier of each fiber}}{\text{yarn count to be spun}}$$

Fibre maturity ratio

$$= \frac{(\text{Normal fibres} - \text{dead fibres})}{200} + 0.7$$

Maturity coefficient

$$= \frac{N + 0.75\,T + 0.45D}{100}$$

where   N = normal fibre
T = thin walled or half mature fibres

Spinning limit

$$= \frac{5315}{\text{Denier of fibre} \times \text{Actual fibres per cross section}}$$

Spinning value of cotton = 78.2Y – 24.8

Where   X = mean fibre length in inches
Y = mean fibre weight/inch (106 oz)

17.15   Formula to predict spinning value from fibre properties

$N = 71.6\,X_1 - 70.8\,X_2 - 20.8\,X_3 + 17.9\,X_4 + 0.037\,X_5 + 4.4\,X_6{}^{14.1}$

Where   N = Highest standard warp count

X_1 = Fibre length (inches)
X_2 = Fibre weight/inch (10^{-6} oz)
X = Ribbon Width (10^{-3} inch)
X_4 = Fibre strength (OZS)
X_5 = Fibre convolutions
X_6 = fibre rigidity (OZ/inch^2 × 10^{-6})]

17.16   Miscellaneous formulae

$$\text{Standard fibre fineness} = \frac{\text{Measured fineness}}{\text{Maturity ratio}}$$

$$\text{Nep count} = \frac{N \times \text{hank of sliver} \times \text{card width in inches}}{0.12 \times 40}$$

or

$$\frac{21 \times N \times \text{hank of sliver} \times \text{card width in inches}}{100}$$

where N-neps/100 sq. in. of web

Cleaning efficiency% of blow room =

$$\frac{\text{Trash in mixing} - \text{Trash in lap}}{\text{Trash in mixing}} \times 100$$

Cleaning efficiency % at card =

$$\frac{\text{Trash in lap} - \text{Trash in sliver}}{\text{Trash in lap}} \times 100$$

Comber fractionation index =

$$\frac{\text{Short fibre \% in comber sliver}}{\text{Short fibre \% in sliver lap}} \times 100$$

$$\text{Diameter of yarn in inches (cotton \& spun silk)} = \frac{1}{0.9 \sqrt{\text{count} \times 840}}$$

$$= \frac{1}{0.9 \sqrt{\text{Yards/lb}}}$$

Threads/inch $= \sqrt{\text{yards/lb}} - 10\%$
$\qquad\qquad = 0.9 \sqrt{\text{yards/ lb}}$
$\qquad\qquad = 26.1 \sqrt{\text{counts}}$

$$\text{Diameter of yarn in inches (worsted)} = \frac{1}{0.9 \ \sqrt{\text{yards/lb}}}$$

$$\text{Nep index} = \frac{\text{Neps per grain in lap} - \text{neps per grain in sliver}}{\text{Neps per grain in lap}}$$

$$\text{Yarn strength index (YSI)} = \frac{\text{Corrected standard sequence yarn CLSP}}{\text{Corrected shirley miniature CLSP}}$$

Where CLSP – count lea strength product

Yarn breaking strength (carded yarn)

$$= 1600 \ (1 + 0.11 \times + 0.01 \ y)$$

where x – difference in staple in 1/16 inch, above or below 1 inch
+ – sign where staple is more than 1 inch
– – sign when staple is below 1 inch
y – difference in count above or below 28^s
+ – sign when it is above 28^s
– – sign when it is below 28^s

The irregularity of random arrangement of fibres is given by the formula

$$U = \frac{85\%}{\sqrt{F}}$$

Where, U – arithmetic mean of all deviations from the average thickness
F – average number of fibres per cross section

Relation between CV and U

$$CV = 5/4 \ U$$
$$U = 4/5 \ CV$$

Formula for yarn strength prediction

$$S_2 = \frac{N_1 S_1 - (N_2 - N_1) \ C}{N_2}$$

Where N_1 is known count
S_1 is known strength
N_2 is new count
S_2 is new strength
C is a constant the value of which is $35 + 0.7 \ (24 - N_2)$
– sign when $N_2 > 24$
+ sign when $N_2 < 24$

The above formula is used for yarn strength prediction of a particular count when the strength of another count from some cotton is known.

Expected least PMD = 0.656 (count × fibre fineness)
PMD – percent mean deviation

$$\text{Degree of perfection} = \frac{\text{Expected PMD}}{\text{Actual PMD}} \times 100$$

Pressley strength conversion

At 0 in. gauge, grams per Tex   = 5.308 × PI_0
At 1/8 in. gauge, grams per Tex = 6.759 × $PI_{1/8}$
PI   = Pressley index

$$\text{Grams per tex} = \frac{\text{lbs per square inch}}{2161}$$

Calculations in weaving preparatory

18.1 Introduction

The Weaving preparatory process is the intermediate process between spinning and weaving. It consists of preparation of the spun yarn for warp and weft yarns to be used in weaving. This chapter gives and introduction to the warp and weft preparatory processes and also deals with the important formulae used in these processes.

18.2 Winding

The object of winding is to remove the faults in the spun yarn and also to convert it into a package suitable for the next process. There are two types of winding, namely, warp and weft or pirn winding. The warp winding produces a bigger package than the ring cop and it may be in the form of a cheese or a cone. The weft winding converts the yarn into a smaller package compared to the ring cop. This can be conveniently used in weaving.

18.2.1 Various formulae in cone winding

Yards wound per mm (calculated)

= rpm of winding drum × circumference of roller

= rpm of winding drum × diameter of winding drum in inches × $\dfrac{\pi}{36}$

Yards wound per mm (actual)

= calculated value × efficiency

= $\dfrac{\text{rpm of roller} \times \text{diameter of winding drum in inches} \times \pi \times \text{efficiency}}{36}$

winding time required in hours

= $\dfrac{K \, (\text{lbs})}{\text{actual production in lbs} \times \text{number of drums used hours}}$

or $\dfrac{K \times \text{count of the yarn}}{\text{production (actual)} \times \text{number of drums used in hanks/hour/drum}}$

K is the quantity of yarn to be wound.

Number of drums required to wind a quality of yarn of known count within a certain time is given by

$$\dfrac{K \text{ (lbs)}}{\text{Time in hours} \times \text{production (actual) in lbs per drum per hour}}$$

or $\dfrac{K \times \text{count of yarn}}{\text{Time in hours} \times \text{production actual in hanks per drum per hour}}$

Number of drums required per warping machine is given by

$$\dfrac{\text{Production of warping machine} + 1\% \text{ for waste etc}}{\text{Production per winding drum}}$$

Total length of yarn warped

$$= \text{length of warp wound} \times \text{number of ends}$$

Total length of yarn to be supplied by the spooler

$$= \text{total length warped} + \text{waste}$$

Number of winding drums required

$$= \dfrac{\text{Total length of yarn required by warping machine}}{\text{Production (actual)/ winding drum}}$$

Actual production in hanks/hour

$$= \dfrac{\text{Rate of winding (calculated) in yards/min} \times 60 \times \text{efficiency}}{840}$$

Actual production in lbs/h $= \dfrac{\text{Actual production in hanks/ hour}}{\text{count of the yarn wound}}$

Efficiency $= \dfrac{\text{Actual production}}{\text{Calculated production}}$

Clearing efficiency

$$= \dfrac{\text{No. of objectionable thick faults removed by slub catcher}}{\text{Total objectionable thick faults present in the yarn before winding}}$$

Knot factor $= \dfrac{\text{Total yarn breaks during winding}}{\text{Yarn breaks due to objectionable faults}}$

18.2.2 Technical specifications of an automatic winding machine (Autoconer)

Winding speed	1000 m/in.
Cop weight-net	75 g
Length of back wind	60 cm
Package weight	1800 g
End breaks per bobbin	1.50
No. of winding heads per machine	60
Splicing time	7.5 s
Traverse length	6 in.
No. of splices per spindle per minute	0.54
Theoretical production	0.975 kg/spindle/h
Efficiency	90%
Actual production	0.87 kg/spindle/h

18.2.3 Drum winding machines

Speed of machine in rpm =

$$\frac{\text{Diameter of line shaft drum}}{\text{Diameter of machine pulley}} \times \text{rpm of line shaft}$$

Calculated length of yarn wound/mm in yards =

Surface speed of drum per min

$$= \frac{\text{Surface speed of drum in inches} \times 22 \times \text{rpm of drum}}{7 \times 36}$$

18.3   Important formulae in pirn winding

$$\text{Count} = \frac{\text{Length in hanks}}{\text{Weight in lbs}}$$

$$\text{Weight of yarn on each cone} = \frac{\text{Length of yarn in hanks}}{\text{Count of yarn}}$$

Time required to wind a quality of yarn of known count

$$= \frac{\text{Quantity in hanks to be wound}}{\text{Production (actual) in hanks/hour/spindle} \times \text{No. of spindles used}}$$

$$\text{Or} = \frac{\text{Quantity in lbs to be wound}}{\text{Production (actual) in lbs/hour/spindle} \times \text{No. of spindles used}}$$

Length of yarn in hanks/hour to be wound

$$= \frac{\text{length of yarn in hanks in the cloth + waste}}{\text{Given time in hours}}$$

$$= \frac{\text{Length of yarn in hanks/hour to be wound}}{\text{Production (actual) in hanks/spindle/hour}}$$

$$= \frac{\text{Length of yarn in hanks in the cloth + waste}}{\text{Given time in hours} \times \text{Production (actual) in hanks/spindle/ hour}}$$

Quantity of yarn in lbs/hours to be wound

$$= \frac{\text{Quantity of yarn in lbs in cloth + waste}}{\text{Given time in hours}}$$

Number of spindles required

$$= \frac{\text{Quantity of yarn in lbs/hour to be wound}}{\text{Production (actual) in lbs/spindle/ hour}}$$

$$= \frac{\text{Quantity of yarn in lbs in the cloth + waste}}{\text{Given time in hours} \times \text{Production (actual) in lbs/spindle/hour}}$$

Number of spindles required for supplying weft to a loom

= Total quantity of weft in hanks required/hour including waste
  × Production (actual)/spindle/hour in hank
= Average quantity of weft in hanks including waste/
  loom/hour
× Production(actual)/spindle/hour in hanks
= Average quantity of weft in hanks in the cloth produced/
  loom/hour + waste

Number of spindles required when the quantity of weft required is given in lbs

$$= \frac{\text{Total quantity of weft in lbs required/loom/hour including waste}}{\text{Production actual in lbs/spindle/ hour}}$$

$$= \frac{\text{Average quantity of weft in lbs including waste/}}{\text{loom/hour} \times \text{no. of looms}}{\text{Production actual in lbs/spindle/hour}}$$

Average quantity of weft in lbs in the cloth produced/loom/hour in lbs +

$$= \frac{\begin{array}{c}\text{Average quantity of weft in lbs in the cloth produced/loom/}\\ \text{hour in lbs} + \text{Waste/loom/hour} \times \text{number of looms}\end{array}}{\text{Production (actual) in lbs/spindle/hour}}$$

$$\text{Efficiency} = \frac{\text{Actual production in yards per hour}}{\text{Calculated production in yards per hour}}$$

$$\text{Actual production in lbs/hour} = \frac{\text{Production (actual) in hanks/ hour}}{\text{count of yarn wound}}$$

18.3.1 Technical specifications of modern high speed fully automatic pirn winder

Pirn speed	12,000 rpm (maximum attainable)
Yarn speed	800 meters (max)
Weight of yarn on a supply cone	1.25 kg (30 tex yarn)
Weight of yarn on a full pirn	57 g
Length of a pirn	203 mm
Diameter of a full pirn	31 mm
Capacity of hopper	1000 empty pirns
No. of spindles/unit	6
Production per operative	315 kg
Power required per unit	1.5 hp
Floor space required for a unit	2186 mm × 890 mm

18.3.2 Efficiency of an automatic pirn winding machine

The efficiency of a fully automatic pirn winding machine may be as high as 95%. This is due to the fact that most of me winder's work is eliminated by the various automatic devices provided on these machines. It is to be noted that lesser the number of spindles per operative higher is the efficiency, other conditions remaining constant. The quality of yarn also affects the efficiency. An inferior quality of yarn will cause more yarn breakages and consequent stoppage of the machine and hence lower the efficiency. The efficiency is also affected by the size and form of supply packages and the size of the pirns produced.

18.4   Warping

The object of warping is to convert the wound packages in the form of cheeses or cones into a bigger package known as the beam so as to make it suitable for the next process of sizing. During this process imperfections such as thick and thin places which were not removed during winding will be removed here.

There are two types of warping, namely

(a)   Direct or beam warping, and
(b)   Sectional warping

In the direct warping the number of ends from cheeses or cones are directly wound on to a warper beam. This operation is known as beaming. The direct warping is immediately followed by sizing. This method is suitable for single yarns that require additional strength by sizing.

In the sectional warping, as the name implies, the yarn is wound in sections on a cylinder with a taper at one end. This operation is called warping and is followed by beaming where the yarn wound in sections is wound on to a weavers beam. It is to be noted that sectional warping eliminates the next operation of sizing. The sectional warping is suitable for doubled and multi coloured warps.

18.4.1 Various formulae in beam warping

$$\text{Beam count} = \frac{\text{Total length of yarn in hanks}}{\text{Weight of yarn in lbs}}$$

$$= \frac{\text{Length of warp in yards} \times \text{Number of ends in the warp}}{840 \times \text{Weight in lbs}}$$

$$\text{Beam count} = \frac{\text{Total length of yarn on the beam in hanks}}{\text{Weight of warp in lbs}}$$

$$= \frac{\text{Total number of ends in the warp} \times \text{Length of warp in yards/840}}{\text{Weight of warp in lbs}}$$

Count of warp yarn

$$= \frac{\text{Total hanks}}{\text{Weight in lbs}}$$

$$= \frac{\text{Length of warp in hanks} \times \text{Number of ends}}{\text{Weight in lbs}}$$

$$= \frac{\text{Length of warp in yards} \times \text{Number of ends}}{840 \times \text{Weight in lbs}}$$

Weight in lbs

$$= \frac{\text{Total hanks}}{\text{Count}}$$

$$= \frac{\text{Length of warp in hanks} \times \text{Number of ends}}{\text{Count}}$$

$$= \frac{\text{Length of warp in yards} \times \text{Number of ends}}{840 \times \text{Count}}$$

Weight of the beam

= Weight of the beam with warp – Weight of the empty beam

Length of warp

$$= \frac{\text{Count} \times \text{Weight in lbs hanks}}{\text{Number of ends}}$$

$$= \frac{\text{Count} \times \text{Weight in lbs} \times 840 \text{ yards}}{\text{Number of ends}}$$

Total length of yarn in warp

= Length of warp × Number of ends

Total length of yarn in warp

= Length of warp × Number of ends

Number of ends in warp

$$= \frac{\text{Count} \times \text{Weight in lbs}}{\text{Length of Warp in hanks}}$$

$$= \frac{\text{Count} \times \text{Weight in lbs} \times 840}{\text{Length of Warp in yards}}$$

Length of warp wound per hour

= Surface speed of wooden drum/minute × 60

= rpm of drum × circumference of drum in inches × 60 × 3.14

Length of warp wound / 8 hours

= Length in yards per mm × 60 × 8 × efficiency

Weight of the beam

= Weight of the beam with warp – Weight of the empty beam

Speed of beam warping machine in rpm

$$= \frac{\text{Diameter of line shaft drum} \times \text{rpm of line shaft}}{\text{Diameter of machine pulley}}$$

Actual production

= Calculated production × Efficiency

Weight of yarn wound per 8 hours

$$= \frac{\text{Total hanks wound}}{\text{Count}}$$

Time required preparing a set of back beams

$$= \frac{\text{Total length of warp to be produced in yards}}{\text{Production (actual) in yards/hours} \times \text{Number of machines used}}$$

Total length of warp length of warp/beam × Number of beams/ set × Number of sets

Therefore time required in hours

$$= \frac{\text{Length of warp/beam} \times \text{Number of beams/set} \times \text{Number of sets}}{\text{Production in yards/ hour (calculated)} \times \text{Efficiency} \times \text{Number of machines used}}$$

Number of warpers per slasher

= Actual production in yards/slasher/hour × average

$$= \frac{\text{Number of ends in the loom beam}}{\text{Actual production in yards/ warper/ hour} \times \text{Average number of ends in a warpers beam}}$$

Production (actual) of warp in yards/hour

= Production (calculated) in yards per hour × Efficiency

Length of yarn in hank warped offends (calculated)/hour

$$= \frac{\text{Production (calculated)} \times \text{Number of ends in the warp}}{840}$$

Length of yarn in hanks warped (actual/hour)

$$= \frac{\begin{array}{c}\text{Production (actual) in yards of warp/hour} \times \\ \text{Number of ends in the warp}\end{array}}{840}$$

$$= \frac{\begin{array}{c}\text{Production (calculated) in yards of warp/hour} \times \\ \text{Number of ends in the warp} \times \text{Efficiency}\end{array}}{840}$$

Weight (calculated) of warp or yarn in lbs warped per hour

$$= \frac{\begin{array}{c}\text{Length of warp in yards warped (calculated)/hour} \times \\ \text{Number of ends in the warp}\end{array}}{840 \times \text{Count of the yarn}}$$

Weight (actual) of warp or yarn in lbs warped per hour

$$= \frac{\begin{array}{c}\text{Length of yarn in hanks warped (actual)/hour} \times \\ \text{Number of ends in the warp}\end{array}}{\text{Count of the yarn}}$$

$$= \frac{\begin{array}{c}\text{Length of yarn in banks warped (calculated)/hour} \times \\ \text{Number of ends in the warp} \times \text{Efficiency}\end{array}}{\text{Count of the yarn}}$$

Number of beams produced (actual)/8 hours

$$= \frac{\begin{array}{c}\text{Length of warp in yards (actual) warped/hour} \times 8 \times 840 \times \\ \text{Length of warp in yards/beam}\end{array}}{840 \times \text{Length of warp in yards/beam}}$$

$$= \frac{\text{Length of warp (calculated) in yards warped/hour} \times 8 \times \text{Efficiency}}{840 \times \text{Length of warp in yards/beam}}$$

Number of warping machines required to wind a known length of warp in a given time

$$= \frac{\text{Total length in yards to be warped}}{\text{Given time in hours} \times \text{Production (actual) in yards/warp per hour}}$$

Number of slashers/beam warper

$$= \frac{\text{Actual production in yards/hour/warper} \times \text{Number of ends warped on each beam warper}}{\text{Actual production in yards/hour/slasher} \times \text{Number of ends slashed on the slasher}}$$

18.4.2 Technical specifications of modern beam warping machine

Size of warping beam	– 1016 mm (flange diameter)
Rate of winding	– 825 meters/mm
Creel capacity	– 1200 cones – 600 active
600 reserve	
No. of spools on the trident	– 15
Length of yarn on a warping beam of 25 tex yarn	– 36,500 meters (approx)
Weight of warp on a warping beam of 25 tex yarn	– 455 kg or more
Weight of yarn on a spool	– 2.72 kg. (max)
Length of 25 tex yarn on a spool	– 1,10,000 meters (approx)
No. of loom beams per warping/section beam	– 20 to 40
Drive	– Individual motor

18.4.3 Various formulae in sectional warping

Number of sections to make warp

$$= \frac{\text{Total number of ends}}{\text{Creel capacity}}$$

Width of a Section

$$= \frac{\text{Width of warp on loom beam}}{\text{Number of Sections}}$$

Number of ends in each section

$$= \frac{\text{Number of ends}}{\text{Number of sections}}$$

Total number of patterns

$$= \frac{\text{Total number of ends} - (\text{Number of extra pattern ends at both ends near selvedges} + \text{total number of selvedge ends})}{\text{No. of ends/pattern}}$$

18.4.4 Technical specifications of a vertical mill warper

Circumference of the mill	18 yards
Height of the mill	10 feet
Creel capacity	250 to 300 bobbins
Ends of warp worked	500 to 3000
Length of warp wrapped	Up to 1000 yards
Speed of the machine	40 rpm
Power required for the machine	1.25 hp

18.5   Sizing

The objective of sizing is to coat the warp yarn with size paste. This helps the yarn to withstand the weaving tensions, by imparting additional strength. The sizing also helps to cover the protruding fibres in the yarn thereby preventing yarn breakage in weaving due to entanglement of neighboring ends. Sizing also increases the abrasion resistance of the yarns.

18.5.1 Material calculations

Weight of size in warp in lbs

$$= \frac{\text{Length in warp yards} \times \text{No. of ends}}{840 \times \text{Counts}}$$

Percentage of size on warp

$$= \frac{\text{Weight of size} \times 100}{\text{Weight of unsized warp}}$$

Weight of sized warp

$$= \text{Weight of unsized warp} + \text{Weight of size}$$

Or   Weight of unsized warp $\times \dfrac{100 + \% \text{ of size}}{100}$

Weight of size to be put on warp

$$= \text{Weight of unsized warp} \times \% \text{ of size required to be put on warp}$$

Count of sized yarn

$$= \frac{\text{No. of ends in warp} \times \text{Length of warp in yards}}{\text{Weight of sized warp in lbs} \times 840}$$

Or   Count of unsized yarn $\times \dfrac{100}{100 + \% \text{ of size on the warp}}$

18.5.2 Production and related calculations

Speed of machine pulley in rpm

$$= \frac{\text{Diameter of line shaft drum} \times \text{rpm of the line shaft}}{\text{Diameter of the machine pulley}}$$

Calculated production of warp length/min

$$= \frac{\text{Circumference of drawing roller in inches} \times \text{rpm of drawing roller}}{36}$$

Calculated production of warp length/8 hours

$$= \frac{\text{Circumference of drawing roller in inches} \times \text{rpm of drawing roller} \times 60 \times 8}{36}$$

Calculated length of yarn produced/8 hours

$$= \frac{\text{Number of ends} \times \text{Yards of warp/mm} \times 60 \times 8}{840}$$

Actual production of warp length / 8 hours

$$= \text{Calculated production} \times \text{Efficiency}$$

Time required sizing a set of beams, in hours

$$= \frac{(\text{Total length of warp in yards} - 40 \text{ yards})}{\text{Production (actual) of sizing machine in yards/hour}} \times \frac{(100 + 1)}{100}$$

Production (actual) of sizing machine, in yards/hour
Number of looms that could be fed by a slasher sizing machine

$$= \frac{\text{Production (actual) of slasher in yards/hour}}{\text{Length in yards of sized yarn consumed by the looms} + \text{Waste}}$$

$$= \frac{\text{Actual production of sized warp in yards/hour}}{\text{Actual production of cloth/hr} + \text{Up-take} + \text{Waste}}$$

$$= \frac{\text{Actual production of sized warp in yards/hr}}{\text{Actual production of cloth/hr}} \times \frac{100 + \% \text{ of regain}}{100}$$

Number of weaver's beams that could be made from a set of back beams

$$= \frac{1}{100} \left\{ \frac{\left(\begin{array}{c} \text{Length of warp in yards on a back beam} - \\ \text{Waste in sizing} \end{array} \right) \times}{100 + \% \text{ elongation in sizing}} \right\}$$

18.5.3 Technical specifications of a modern sizing machine

Diameter of drying cylinders	–	800 mm
Working width of cylinders	–	1400 to 2200 mm
Width of cylinders	–	1600 to 2400 mm
No. of cylinders	–	3, 5, 7, 9, or 11
Speed of the warp on the slasher	–	100 to 150 meters/min
Creep speed	–	1.8 to 2 meters/min
Length of drier with 9 drying cylinders	–	4600 mm
Steam pressure	–	3 to 4 atmospheres
Winding tension	–	1000 kg and above
Squeezing pressure of squeezing roller	–	Up to 970 kg
Water evaporation/hour in kg for 9 cylinder slasher	–	200 to 950 kg.
Diameter of loom beam	–	900 mm (flange (diameter)
Power required	–	5 to 10 kilowatts

18.6   Drawing in

The object of drawing in process is to make the sized beam ready for weaving by drawing the individual warp ends through heald and dents of the reed.

18.6.1 Details relating to specifications of various reeds

Table 18.1 explains the specifications of various systems of reed numbering and useful for reed calculations:

Table 18.1 System based on the no. of dents in a given space

Name of the system	Basis of numbering
Stock port	Number of dents/2 inches
Radcliff	Number of dents per inch
Hudderfield	Number of dents per inch
Metric	Number of dents/decimeter
Bolten	20 dents per 24.5 inches
Bradford	20 dents per 36 inches
Blackburn	20 dents per 45 inches
Irish	100 dents per 40 inches
Leeds	19 dents per 9 inches
Maccles field	100 dents per 36 inches

18.6.2 Reed and heald calculations

Formula for conversion of reed number from one system to another

Required reed count

$$= \frac{\text{Dents per inch in known system}}{\text{Dents per inch in required system}} \times \text{Reed count in known system}$$

$$\text{Count of heads required for a weave} = \frac{\text{Number of ends/inch}}{\text{Number of heald shafts}}$$

18.7 Silk reeling and throwing

18.7.1 Introduction

Silk reeling is basically a process of unwinding the filament silk from the cocoon. This is achieved by cooking the cocoons at a particular temperature in a water bath so as to remove the gummy matter and then unwinding the continuous filament from the cocoon. Silk reeling is an area which needs to be developed in India. In India, there are three distinct reeling devices in mulberry sector namely.

(a) Charka
(b) Cottage basin system which is an improvement over charka, and
(c) Multi end basin system which is supposed to be the modern version in India

18.7.2 Charka

The charka reeling system is an Italian version or floating system of reeling. It improves the reliability of inferior and defective cocoons without considering the quality of silk. The charka, which is a simple device consists of a large cooking cum reeling pan where boiling water is kept. The cocoons are cooked in it and filaments collected in a bunch after brushing are passed through a hole on an ordinary thread guide device. Later, the thread is crossed with another co-thread for forming a chambon-type croissure in order to agglutinate the filaments and remove the water from the body of' the thread. Then it is passed through a distributor before it is wound on to a large wooden reel. Four threads are maintained in this device. One person rotates the reel by hand and another person sitting near the cooking pan manipulates the cocoon cooking and reeling.

18.7.3 The cottage basin

The cottage basin is a reeling device, which is an improved version over charka and it is indigenously designed on the principle of Japanese multi end reeling machine. Here cocoon cooking is done separately in a boiling water basin and reeling is done in a hot water basin. Each basin has six ends and each thread is first passed through a button to reel. So, the quality of silk is superior to charka silk. Superior quality cocoons like bivoltine can be reeled on this device. But, cottage basin is generally hand driven and alignment of the basin is not sufficiently perfect. As a result, production of superior quality silk conforming to international standard is not possible.

18.7.4 Multi-end reeling basin

This reeling device is a further improved version over the cottage basin and it is power driven. Boilers are installed and steam is used for cooking and reeling purpose and also for cocoon stifling in the special steam chambers. Since recently, however, hot air drying methods are being introduced for bivoltine cocoons in some of the filature units in the country for improving the working efficiency and quality of reeled silk.

Normally, cocoon cooking is done according to single pan system as in the case of cottage basin. In multi end basins, there are some additional attachments such as Jetteboute which picks up the filaments to increase the efficiency of cocoon feeding. The distribution system is further improved and individual brake motion for each reel is also provided so that the overall working efficiency of the basin is enhanced. Normally, each basin consists of 10 ends.

In principle, the multi end reeling machine is supposed to be a modern reeling device in India and it is possible to use superior quality cocoons like bivoltines on these machines with better performance.

18.7.5 Production calculation

The reeling production is calculated as follows:

$$\frac{\text{Reeling speed (rpm)} \times \text{Circumference of reel (m)} \times 8 \times 60 \times}{\text{No. of ends} \times \text{denier}}$$

9000

Example: A multi-end silk reeling machine has the following particulars:

(i)   Circumference of the reel – 0.7 m
(ii) Denier of the silk being reeled – 22
(iii) No. of ends – 10
(iv) Reeling speed – 150 rpm

Calculate the production and efficiency.

$$\frac{150 \times 0.7 \times 8 \times 60 \times 10 \times 22}{9000} = 1232 \text{ g}$$

$$\text{Therefore efficiency } \% = \frac{1000}{1232} \times 100 = 81.2\%$$

18.7.6 Silk throwing

This is basically a preparatory process prior to weaving. Before the raw silk is woven into a fabric, it must go through a series of operations which conditions it for the loom. The series of preliminary preparatory processes involved are as follows:

(i)   Winding
(ii) Doubling
(iii) Twisting
(iv) Re-winding
(v)   Warping
(vi) Pirn winding

(i) Winding

The main functions of winding are to put the yarn in a long continuous length to suit later processes and also to eliminate imperfections such as slubs, weak places, dirt and so on.

Winding is necessary to rewind silk yarn from hank onto bobbins. During this winding process, hard gum spots, loose ends, defective knots etc. are removed. Hence, the quality of silk thread in the bobbin is comparatively better than that of the hank. Winding should be done at 27° ± 2°C temperature and 65% ± 2% relative humidity. If brakes are less during winding, it indicates that the quality of silk is good. There is a standard breakage rate for different grades; for example, winding breaks should not be more than 12 for 40 bobbins per hour for A grade silk (20/22 d).

The winding process primarily indicates the weaving efficiency and quality of fabric, since the rate of winding breaks and knotting have impact on weaving efficiency and quality of fabric, since the rate of winding breaks and knotting have impact on weaving efficiency and quality of fabric. Generally for finer raw silk, winding speed should be lower and vice versa. Production per day per drum of winding machine is 100–150 g depending upon the denier and quality of raw silk.

(ii) Doubling

The object of doubling is to double the individual threads. Doubling avoids unevenness and the strength of doubled yarn is correspondingly better than the single thread. The doubling machine is more or less like winding machine. It has a capacity of 50–100 bobbins and it is double sided. Its production capacity will be in the range of 2–4 times that of winding machine depending upon the number of ply.

(iii) Twisting

The twisting of silk is based on the up twister principle. There is a vertical spindle on which doubling bobbin is mounted and yarn from this is wound on to a perforated bobbin mounted horizontally and driven by surface contact. Twist is imparted on account of difference between the speed of the spindle and the winding drum. The direction of twist has impact on cover of fabric. The production is affected by twist rate. Higher the twist rate, lower is the production. Normally, production per spindle per day is 25 to 30 g (two ply) for 20/22 denier yarn. Twisting affects the brilliancy of the yarn. This is because, the roughness of the thread's surface caused by twisting and the ridges of the spirals causing shadows with loss of reflected light. As the number of twist increases, the brightness will be subdued more.

(iv) Re-winding

The rewinding machine is practically like winding machine. Its production capacity is more, since normally double yarn is wound on this. If two ply yarn is rewound, production rate would be more than two times as compared to winding machine. Throwing process is carried in standard atmospheric conditions, in order to reduce the rate of breakages, so that the production and quality of yarn arc maintained.

   Presently the modern trend of processing is the use of single yarn in weaving, thus the throwing process may be omitted. Instead of re-reeling raw silk from reeling swift, it is wound on to cheese form. These cones or cheeses are directly used in warping and pirn winding. Content of yarn in cones or cheeses is more, so, efficiency of warping and pirn winding would be high, besides reduction in cost of manufacture on account of avoiding winding etc. However for special yarns like crepes, etc. throwing process is very much necessary. Waste produced during throwing is 1 to 2 per cent.

(v) Warping

In silk weaving, normally sectional warping is followed because of the fine denier of silk thread and consequently higher number of ends required. Warping machine mainly consists of two parts, namely,

(i)   warping creel, and

(ii) warping drum

The type and arrangement of creel depends upon the system of weaving, investment, space available, etc.

(vi) Pirn winding

Pirn winding is necessary to prepare weft yarn. Generally, the pirns used for power looms are larger and the yarn content will also be more. Thus the pirn winding machine is generally used for power loom weaving. Pirn winding machine may be automatic or non automatic. The automatic pirn winding machines have certain devices such as tension control, bunching mechanism, coil distributor and thread stop motion. Thus, the package of pirn will be perfect and content of yarn is more, so that weft replenishment will be at lower frequency, thereby increasing the weaving efficiency. The handloom pirn is smaller and thus, hand operated charka is used for preparing the pirns.

Generally, the type of yarn preparation depends on the type of fabric that is being woven. Chiffon, georgette and crepe need extensive yarn preparation. The machinery used must be capable of running at higher speeds and yarn clearers must be used on these machines. Yarn content on the packages must be increased so as to facilitate larger lengths, of yarn for warping.

Calculations in the weaving process

19.1 Introduction

This part of the chapter is concerned with the speed, production and efficiency. The speed of a plain loom depends on the reed space. In case of other types of looms such as dobby, jacquard and drop box looms other factors in addition to reed space will have to be considered. Hence the speed of these looms is generally lesser than a plain loom.

19.2 Speed and pick calculations

Speed of a loom

rpm of crank shaft

$$= \frac{\text{Diameter of line shaft drum} \times \text{rpm of the shaft}}{\text{Diameter of loom pulley}}$$

rpm of bottom shaft

$$= \text{rpm of crank shaft} \times \frac{1}{2}$$

Picks per inch

$$= \frac{\text{Number of picks put in}}{\text{Length of cloth drawn forward in inches}}$$

Picks per inch in the cloth

$$= \frac{\text{Constant} + 1\frac{1}{2}}{\text{Change wheel}}$$

19.3 Take up motion

The object of this motion is to draw the cloth forward at a required rate as each pick is inserted and also to influence the picks/inch inserted. This motion is of the positive and negative type. The positive type consists of 5 wheel and 7 wheel take up motions.

19.3.1 Five wheel take up motion

Figure 19.1 below shows a five wheel take up motion. For every pick the ratchet wheel E is moved one tooth by the catch D, which in turn gets movement from the sley sword through A. The motion is transferred to the cloth roller J, through a train of wheels E, C, G, H and I. The rate at which the cloth is drawn forward depends on the size of the change wheel C, which is changed, whenever the picks per inch are required to be changed and hence its name.

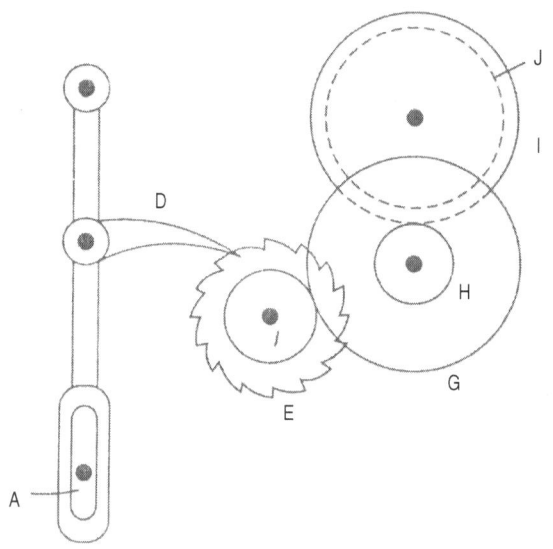

19.1 Five wheel take up motion.

$$\text{Picks/inch} = \frac{\text{No. of picks put in}}{\text{Length of cloth drawn forward in inches}}$$

If E represents the number of teeth of the ratchet wheel then the number of picks per revolution of the ratchet wheel = No. of teeth of ratchet wheel E.

Length of cloth drawn forward in inches per revolution of ratchet wheel.

$$= \frac{C}{G} \times \frac{H}{I} \times \text{circumference of beam in inches}$$

Therefore, picks per inch

$$= \frac{E \times G \times H}{H} \times \frac{1}{\text{Circumference of beam in inches}}$$

Because of the size of the wheels E, G, I, H and the circumference of the beam are kept constant in any particular loom. When the picks per inch in cloth are required to be changed, it is clone by changing the change wheel. There is shrinkage in length when the cloth is taken out of the loom because the cloth will be relieved of the tension under which it was held on the loom. Therefore actual picks per inch in the cloth will be higher. The amount of shrinkage varies generally from 1% to 5%. Under normal circumstances it may be taken as 1.5%.

$$\text{Therefore picks/inch in the cloth} = \frac{\text{Constant}}{\text{Change wheel}} + 15\%$$

$$= \frac{\text{Constant } K_1}{\text{Change wheel}}$$

$$\text{and the change wheel} = \frac{\text{Constant } K_1}{\text{Picks per inch in cloth}}$$

The constant K_1 is called the dividend.

Particulars of several combinations of gearings which are used for 5-wheel take-up motion together with the corresponding dividend for each such combination of gearings are given below:

Ratchet wheel	60	60	60	50	50	50
Stud wheel	100	100	120	100	120	120
Pinion	12	12	15	75	15	18
Beam wheel	60	75	75	75	12	100
Circumference of beam	15"	15"	15"	15"	15"	16"
Dividend	2030	2538	2436	2114	2030	2114

In case of a 5 wheel take up motion the constant is given by

$$\text{Constant (dividend)} = \frac{\text{Ratchet wheel} \times \text{Stud wheel} \times \text{Beam wheel} + \frac{1}{2}}{\text{Pinion} \times \text{Circumference of beam}}$$

19.3.2 Pickles 7 wheel take up motion

Figure 19.2 shows the seven wheel take-up motion.

Unlike the 5 wheel motion explained previously, the change wheel in this motion is a driven one, so the picks per inch in the cloth will increase or decrease with the increase or decrease respectively of the change wheel. "S' is called the standard wheel.

$$\text{Picks/inch} = \frac{\text{No. of picks put in}}{\text{Length of cloth drawn forward in inches}}$$

If *A* represents the number of teeth of the ratchet wheel then,

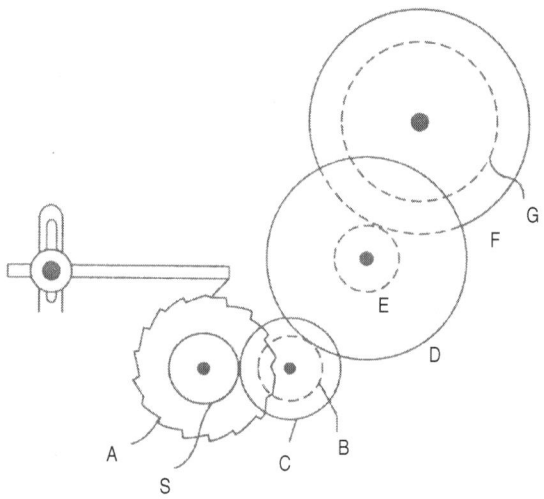

19.2 Seven wheel take up motion.

No. of picks per revolution = A, i.e. No. of teeth on ratchet wheel of ratchet wheel

19.4   Take up calculations

Length of cloth in inches drawn forward per revolution of the ratchet wheel

$$= \frac{\text{Standard wheel} \times \text{swing pinion} \times \text{pinion} \times \text{circumference of beam}}{\text{Change wheel} \times \text{stud wheel} \times \text{beam wheel}}$$

$$= \frac{\text{S} \times \text{B} \times \text{E} \times \text{circumference of beam}}{\text{C} \times \text{D} \times \text{F}}$$

Therefore picks per inch

$$= A \times \frac{C}{S} \times \frac{D}{B} \times \frac{F}{E} \times \frac{1}{\text{Circumference of bean}}$$

$$= C \times \text{constant}$$

$$= C, \text{ when a standard wheel of 36 teeth is used.}$$

Then constant = $\dfrac{89 \times 90 \times 24}{36 \times 15 \times 24 \times 15.05}$ + 1.5% = 1

Picks which a tooth of the change wheel represents will vary according to the size of the standard wheel, which is generally a multiple of 9. The table below shows the value of each tooth of the change wheel corresponding to standard wheels of different sizes.

Size of standard wheel	Value of each tooth of change wheel (picks per inch)	Size of standard wheel	Value of each tooth change wheel
9	4	45	4/5 pick per inch
18	2	54	2/3 pick per inch
27	1–1/3	63	4/7 pick per inch
36	1	72	1/2 pick per inch

In a seven wheel take up motion picks per inch

$$= \frac{\text{Number of picks put in}}{\text{Length of cloth drawn forward in inches}}$$

Length of cloth in inches drawn forward per revolution of the ratchet wheel

$$= \frac{\text{Standard wheel} \times \text{Sewing pinion} \times \text{Pinion} \times \text{Circumference of beam}}{\text{Change wheel} \times \text{Stud wheel} \times \text{Beam wheel}}$$

19.5   Production and related calculations

Average rpm of looms in shed

$$= \frac{\text{Total rpm}}{\text{Total looms}}$$

Average reed space of loom shed

$$= \frac{\text{Total reed space}}{\text{Total looms}}$$

Production in yards of a group of looms at 50 picks/inch

$$= \frac{\text{Sum of the actual production in yards of all the looms in the group} \times \text{actual picks one inch in the cloth}}{50}$$

(The group is considered to be looms weaving cloth with same number of picks/inch)

(Total production of shed in yards at 50 picks/inch)

= (Production of group × in yards at 50 picks/inch) + (production of group Y in yards at 50 picks/inch)

Production (calculated) in yards/hour

$$= \frac{\text{Picks/min} \times 60}{\text{Picks/min} \times 36} = \frac{5 \times \text{Picks/min}}{3 \times \text{Picks/in}} = \frac{1.667 \times \text{Picks/min}}{\text{Picks/in}}$$

Efficiency

$$= \frac{\text{Actual production}}{\text{Calculated production}}$$

Actual production in yards/hour

$$= \frac{(\text{Picks/min}) \times 60 \times \text{Efficiency}}{(\text{Picks/in}) \times 36}$$

$$= \frac{5 \times (\text{Picks/min}) \times \% \text{ of Efficiency}}{3 (\text{Picks/in}) 100}$$

Time required to weave a known length of warp on a weaver's beam

$$= \frac{\text{Length in yards of cloth from the beam}}{\text{Actual production in yards of cloth/hour}}$$

$$= \frac{(\text{Length in yards of warp} - \text{uptake}) - \text{waste of warp in yards}}{\dfrac{\text{Picks/min} \times 60 \times \text{Efficiency}}{\text{Picks/in the cloth} \times 36}}$$

Length of yarn on bobbin in case

$$= \frac{\text{Count} \times \text{weight of yarn in ozs} \times 840 \times 36}{16}$$

Therefore length of cloth per bobbin

$$= \frac{\text{Length of yarn on bobbin in inches}}{\text{Picks/inch} \times \text{reed width in inches} \times 16}$$

$$= \frac{\text{Count} \times \text{weight of yarn in ozs} \times 840 \times 36 \times \text{inches}}{\text{Picks/inch} \times \text{reed width in inches} \times 16}$$

$$= \frac{1890 \times \text{Count} \times \text{weight of yarn in ozs yards}}{\text{Picks/inch} \times \text{reed width in inches} \times 16}$$

Number of bobbins/ loom/ hour

$$= \frac{\text{Loom production (actual)/hour}}{\text{Actual length of cloth/bobbin}}$$

Loom production in yards/hour

$$= \frac{\text{Picks/min} \times 60 \times \text{efficiency}}{\text{Picks/inch} \times 36}$$

Length of cloth in yards/bobbin

$$= \frac{\text{Length of yarn inches}}{\text{Picks/inch in cloth} \times \text{reed width in inches} \times 36}$$

$$= \frac{\text{Count of weft} \times \text{weight of weft in lbs} \times 840 \times 36}{\text{Picks/inch in cloth} \times \text{reed width in inches} \times 36}$$

$$= \frac{\text{Count of weft} \times \text{weight of weft in ozs} \times 840}{\text{Picks/one inch in cloth} \times \text{reed width in inches} \times 16}$$

Therefore number of bobbins/hour/loom

$$= \frac{\text{Picks/mm} \times 60 \times \text{efficiency})/(\text{picks/inch} \times 36)}{(\text{count of weft} \times \text{weight of weft in oz.} \times 840)/}$$
$$(\text{picks/inch} \times 16 \times \text{reed width in inches})$$

$$= \frac{\text{Picks/mm} \times 60 \times 16 \times \text{reed width in inches} \times \text{efficiency}}{31.5 \times \text{count of weft} \times \text{weight of weft in ozs.}}$$

Alternative formula

Number of bobbins/loom/hour

$$= \frac{\text{Weight of weft in the cloth produced/ hour}}{\text{Weight of weft on bobbin}}$$

Weight of weft in cloth in ozs

$$= \frac{\begin{array}{c}\text{Length of cloth in yards/hour} \times \text{picks/one inch in cloth} \times \\ \text{reed width in inches} \times 16\end{array}}{840 \times \text{count of weft}}$$

Therefore number of bobbins/loom/hour

$$= \frac{\begin{array}{c}\text{Length of cloth in yards/hour} \times \text{picks/inch in cloth} \times \\ \text{reed width in inches} \times 16\end{array}}{840 \times \text{count of weft weight of weft in ozs on a bobbin}}$$

$$= \frac{\text{Length of cloth in yards/hour} \times \text{picks/inch in cloth} \times \text{reed width in inches} \times 16}{840 \times \text{count of weft} \times \text{weight of weft in ozs on a bobbin}}$$

Another method

Number of bobbins/loom/hour

$$= \frac{\text{Length of weft in cloth produced/hr}}{\text{Length of weft (actual) on a pirn}}$$

$$= \frac{\text{Length of cloth in yards/hr} \times \text{picks/ 1 inch in cloth} \times \text{reed width in inches}}{\text{Length of weft (actual) in yards on a pirn}}$$

$$= \frac{\text{Length of cloth in yards/hr} \times \text{picks/1 inch in cloth} \times \text{reed width in inches}}{\text{count of weft} \times \text{weight of yarn in ozs on a bobbin/16}}$$

Time in minutes to exhaust a bobbin on loom

$$= \frac{\text{Length of cloth woven/bobbin}}{\text{Production (actual) in yards/min/ loom}}$$

Length of warp required/loom/hour

= Production (actual)/hour/loom + contraction in weaving + waste

19.6   Production planning

19.6.1 Average rpm or average picks per minute of looms in a weaving shed

These data are sometimes necessary to determine the efficiency and production of a weaving shed having a number of looms. The rpm of individual looms may be measured by direct counting method or by a tachometer. In counting the rpm by direct counting method or with a tachometer, it is better to take the mean of several readings for each loom. To get accurate result it is also necessary to measure the rpm of the same loom at different times and take the mean value as the rpm of that loom.

In a weaving shed there are looms of different widths and it is possible to find the average rpm of different loom widths from the rpm of individual looms. The average picks of the shed can then be calculated either by adding together the picks per minute of the individual looms and dividing the sum by the total number of looms in the shed or by dividing the sum of the products of average picks per minute of looms of each of the sizes and their corresponding number by the total looms in the shed.

Size of looms	Number of looms	Average rpm of looms of different sizes
120 cm	160	260
150 cm	80	240
180 cm	70	220
220 cm	90	190

Example: A weaving shed has 400 looms and is divided into 5 sections of 80 looms per section. Particulars of the looms are as follows:

Calculate the average rpm and average picks per minute of the shed.

Total number of looms   = 160 + 80 + 70 + 90 =   400
Total rpm of 120 cm looms = 160 × 260   = 41600
Total rpm of 150 cm looms = 80 × 240   = 19200
Total rpm of 180 cm looms = 70 × 220   = 15400
Total rpm of 220 cm looms = 90 × 190   = 17100

Total   **93300**

Therefore, average rpm of looms in shed $= \dfrac{\text{Total rpm}}{\text{Total looms}} = \dfrac{93300}{400} = 233.25$

19.6.2 Average reed space

The average reed space may be calculated by adding together the reed space of the individual looms and dividing the sum by the total number of looms in the weaving shed. An alternative method of doing this is to multiply each size by the number of looms it corresponds. The sum of these products is divided by the total number of looms. The sum of these products divided by the total number of looms in the shed will give the required average reed space.

Example: A weaving shed has 360 looms of the following particulars. Find the average reed space of the shed.

Width of looms	Number of looms
110 cm	140
160 cm	70
190 cm	60
250 cm	90

Total number of looms = 140 + 70 + 60 + 90 =   360
Total reed space of 110 cm loom = 110 × 140   = 15400
Total reed space of 160 cm loom = 160 × 70   = 11200
Total reed space of 190 cm loom = 190 × 60   = 11400
Total reed space of 250 cm loom = 250 × 90   = 22500
                                Total   **= 60500**

$$\text{Therefore, average reed space of the shed} = \frac{\text{Total reed space}}{\text{Total looms}} = \frac{60500}{360}$$

$$= 168 \text{ cm}$$

19.7   Factors affecting the production of looms

The loom production is chiefly affctcc1 by the speed of the loom. Other factors which affect production are quality of warp and weft, picks inserted per inch of the cloth, types of loom etc. In shuttle less looms, the loss of production and consequently efficiency of the loom due to its stoppages for replenishing weft have practically been eliminated.

19.8   Loom and weaving shed efficiency

The efficiency of a loom depends on the amount of time lost due to loom stoppages. Less the production loss due to loom stoppages more is the efficiency. In general, the factors on which production depends are also responsible for the efficiency.

$$\text{Efficiency} = \frac{\text{Actual production}}{\text{Calculated production}} \times 100$$

Efficiency may be classified under the following heads:

Running efficiency

It does not take into account such stoppages as for cleaning, etc.

Overall efficiency

It takes into account such stoppages as for cleaning, etc., and so it is lower than the running efficiency.

Shed efficiency

It takes into account the looms stopped say, for want of warps suitable for them for a week or so or for other causes, say for a day or two or more.

Calculations in the knitting process

20.1 Introduction

Knitting is a process of manufacturing a fabric by the intermeshing of loops of yarns. When one loop is drawn through another loop, a 'stitch' is formed. Stitches may be formed in a horizontal or in a vertical direction. There are two basic types of knitting, namely,

(a) Warp knitting, and
(b) Weft knitting

The threads running parallel to the selvedges are called the warp and the threads running at right angle to the selvedges are called weft.

20.2 Warp knitting

Warp knitting is a method of forming a fabric by knitting means in which the loops are made in a vertical way along the length of the fabric from each warp yarn and intermeshing of loops takes place in a fiat form on a length-wise basis. In this method, numerous ends of yarns are being fed simultaneously to individual needles placed in a lateral fashion. Most of the knitted structures come out in flat or open width form, though there are a few warp knit fabrics that are produced in a tubular form.

Figure 20.1 shows a single bar warp knit structure. One shaded series of loops, which run in a zig zag fashion vertically, shows the manner in which an individual thread is intermeshed with other warp threads in the construction of a warp knitted fabric.

20.3 Weft knitting

Weft knitting is a method of forming a fabric by knitting means in which the loops are made in a horizontal way from a single yarn and intermeshing of loops takes place in a circular or flat form on a course-wise basis. In this method, one or more yarns (called 'feeds') are being fed one at a time, to a multiplicity of knitting needles, placed in either lateral or circular

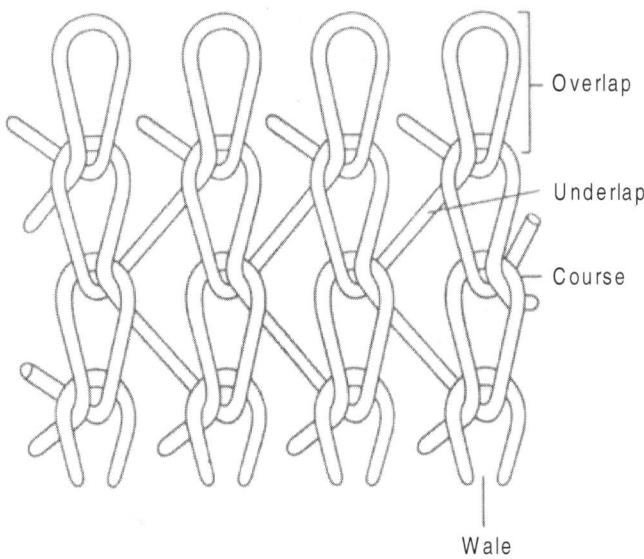

20.1 Warp knitted structure

fashion. Most of the weft knitted structures comes out in a tubular form though a good proportion of them are manufactured in the form of yard goods in open width.

Figure 20.2 shows the 'face' side of a single weft-knit structure. The shaded series of loops are formed in a horizontal or weft direction. Hence the name weft knit structure.

The row of loops in horizontal direction is called 'a course' while the column of loops in vertical direction is called a 'wale'. The 'courses' and 'wales' per unit space determine one of the specifications of a knitted fabric.

20.4   Basic terminologies

Courses

These are rows of loops across the width of the fabric produced by adjacent needles during the same knitting cycle, and are measured in units of courses per centimeter. The number of courses determines the length of the fabric. Figure 20.3 indicates the course made of needle loop and sinker loop.

Wales

A wale is a vertical column of needle loops. The number of wales

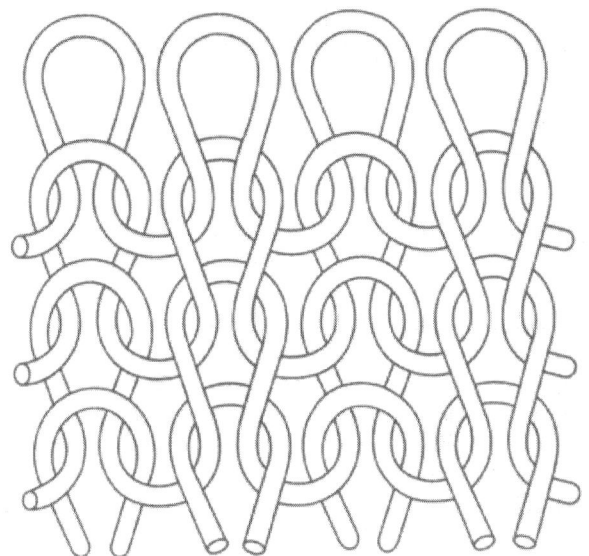

20.2 Weft knitted structure

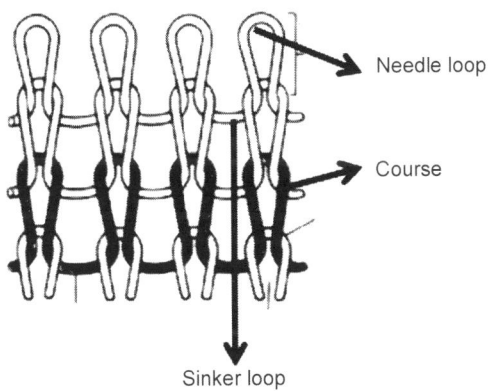

Needle loop

Course

Sinker loop

20.3 Course.

determines the width of the fabric and they are measured in units of wales per centimeter. Figure 20.4 indicates the wales and also the needle loop.

Stitch density

It is the total number of needle loops in a given area.

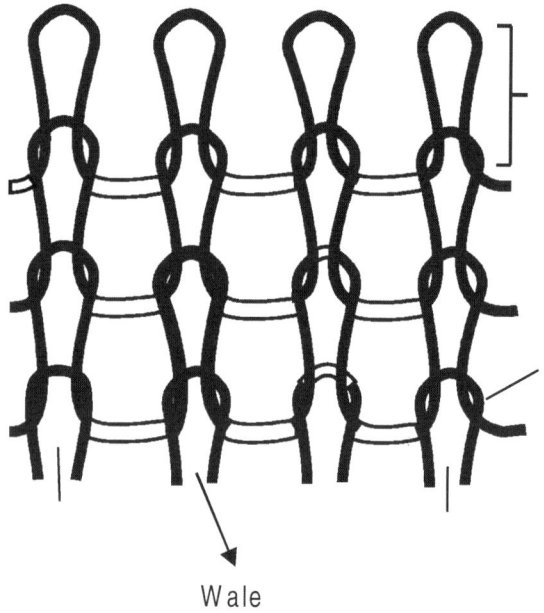

Wale

20.4 Wale

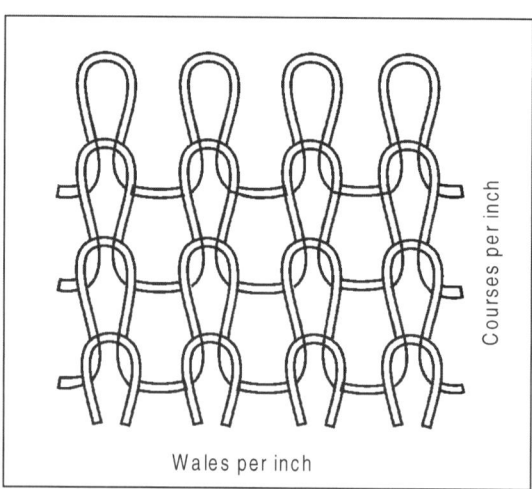

20.5 Stitch density

Stitch length or Loop length

It is the fundamental unit of the weft knitted structure and its length and shape determines the fabric dimensions, which in turn can be affected by the yarn used and the finishing techniques employed, such as heat setting.

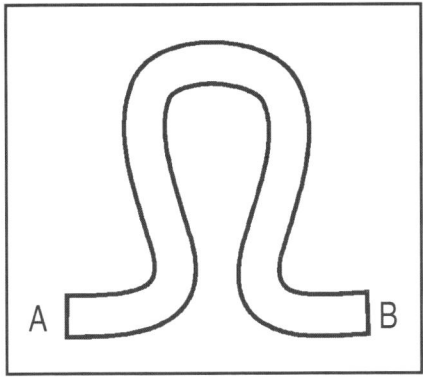

20.6 Stitch length

It is the length of yarn in the knitted loop (From point A to B in Fig. 20.6). Stitch length is one of the most important factors controlling the properties of knitted fabrics and is measured in millimeters.

It can be determined by removing one course length from a fabric and dividing this length by the total number of needles knitting that length of yarn. Generally, the larger the stitch length, the more open and lighter the fabric.

Machine gauge

This is the number of needles per inch in a knitting machine. In the case of a flat bar machine the gauge is 5–14 needles per inch and in circular machines the gauge is 5–40 needles per inch.

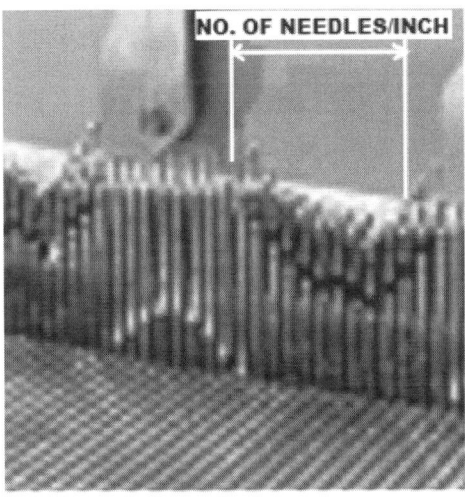

20.7 Machine gauge

Needle

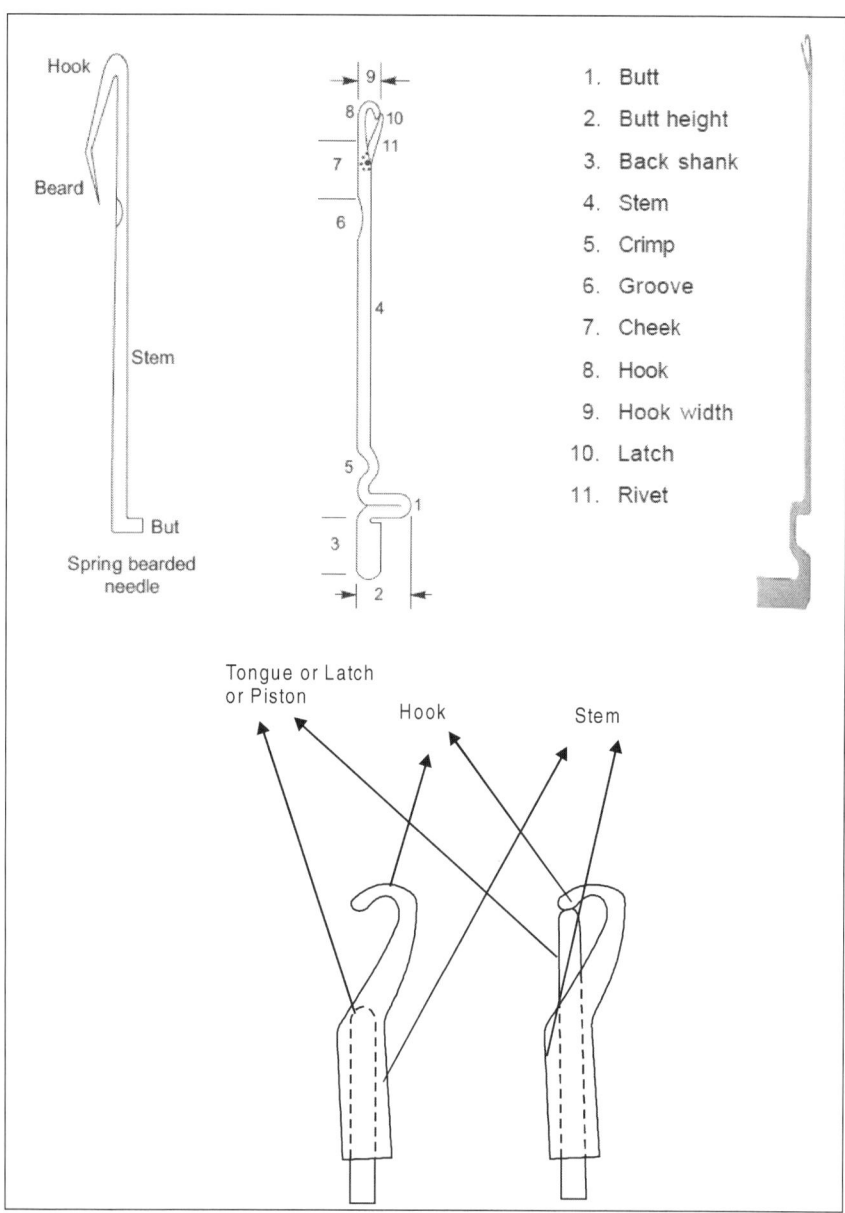

1. Butt
2. Butt height
3. Back shank
4. Stem
5. Crimp
6. Groove
7. Cheek
8. Hook
9. Hook width
10. Latch
11. Rivet

20.8 Spring bearded, latch and compound needles

It is the heart of the knitting machine since it helps in the loop formation. There are three types of needles, namely, the latch needle, compound needle and bearded needle. The latch needle is the most widely used in weft knitting and has the advantage of being self acting, though it is slightly more expensive to produce. The compound needle is increasingly used for warp knitting. It is expensive, with each part requiring separate and precise control. The bearded needle is the simplest and cheapest to manufacture but it does require an additional element to close the beard during knitting. It is used in straight bar machines in weft knitting.

Plain single jersey fabric

Plain single jersey is the simplest weft knitted structure that it is possible to produce on one set of needles.

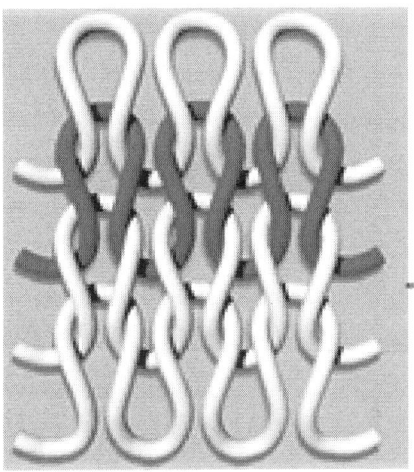

20.9 Plain single jersey

Rib fabric

It is a knitted fabric with vertical rows (wales) of loops meshed in the opposite direction to each other.

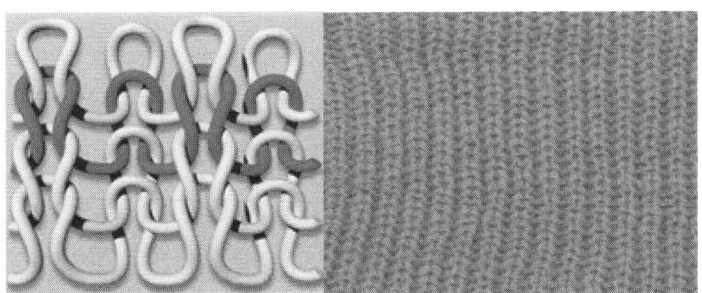

20.10 Rib fabric

Purl fabric

This is also known as link fabric. A 1 × 1 purl fabric has loops knitted to the front and back on alternate courses, in contrast to a 1 × 1 rib fabric which is knitted to the front and back on alternate wales. It resembles the plain single jersey fabric's face side on both of its front and back.

Interlock fabric

It consists of two 1 × 1 rib fabrics knitted one after the other by means of two separate yarns, which knit alternately on the face and back of the fabric and are interlocked together.

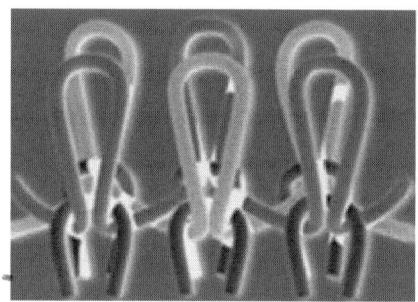

20.11 Interlock fabric

Tuck stitch

It is a stitch made when a needle takes a new loop without clearing the previously formed loop, so that loops are accumulated on the needles.

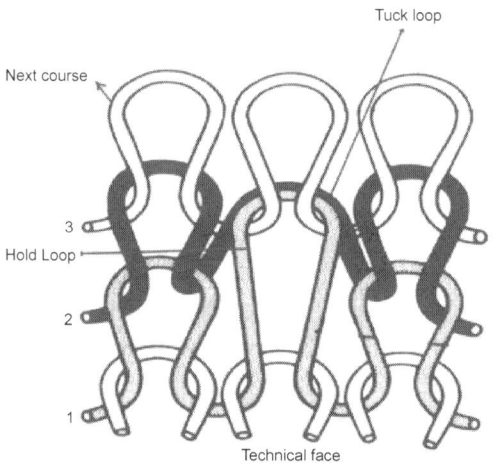

20.12 Tuck stitch

Float or Miss stitch

It is produced by a needle holding the old loop while the two adjacent needles are raised and cleared to produce a new knitted loop.

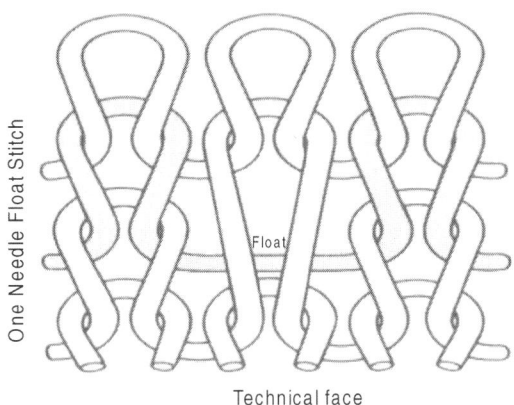

Technical face

20.13 Float stitch

20.5 Calculations pertaining to weft knitting

Relation between yarn number and machine gauge

$$\text{Worsted yarn number} = \frac{(\text{Needles per inch})^2}{\text{Constant}}$$

The values of constants for different stitches are given below.

Plain Jersey – 14
1 × 1 Rib   – 6
2 × 2 Rib   – 8
Interlock   – 10

For flat 1 × 1 rib machine

$$\text{Worsted yarn number} = \frac{(\text{Gauge})^2}{9}$$

For French circular sinker wheel machine

$$\text{Worsted yarn number} = \frac{(\text{Gauge})^2}{28} \qquad \dots (\text{Fine yarns})$$

$$(\text{or}) \qquad = \frac{(\text{Gauge})^2}{47} \qquad \dots (\text{Fine yarns})$$

For calculating the production in weft knitting machine, the following parameters have to be considered.

(a) Number of knitting feeders (N_1)
(b) Number of revolutions per minute (N_2)
(c) Efficiency of the machine in percent (F)
(d) Number of courses per inch (N_3)
(e) Stitch length in inches (L)
(f) Number of needless (N_4)
(g) Yarn number in indirect system (N_5)

Using the above parameters, the following can be calculated:

(i) Yards of knitted fabric per hour $= \dfrac{N_1 \times N_2 \times E \times 60}{N_3 \times 36 \times 100} = \dfrac{N1 \times N2 \times E}{60 \times N3}$

(ii) Production per hour in lb $= \dfrac{N_1 \times N_2 \times E \times L \times N_4 \times 60}{N_5 \times \text{yards per hank} \times 36 \times 100}$

$$= \dfrac{N_1 \times N_2 \times E \times L \times N_4}{N_5 \times 50400}$$

(for cotton-since 840 yards – 1 hank)

(or) $$= \dfrac{N \times N_2 \times E \times L \times N_4}{N_5 \times 33600}$$

(for worsted – since 560 yards – 1 hank)

20.6   Fabric calculations

Weight per linear yard in lbs

$$= \dfrac{\text{Total needles} \times \text{Courses per inch} \times 36}{\text{Stitches per foot} \times 3 \times 840 \times \text{Cotton count}}$$

$$= \dfrac{\text{Total no. of needles}}{\text{Wales per inch}}$$

Weight per square yard in lb

$$= \dfrac{\text{Weight per linear yard} \times 36}{\text{Fabric width in inches}} \quad \text{(for flat fabric)}$$

(or) $$= \dfrac{\text{Weight per linear yard} \times 36}{\text{Fabric width in inches } 2} \quad \text{(for tubular fabric)}$$

Fabric width in inches – Number on needles × waste spacing
= Needles × 4 – yarn diameter per wale

$$\frac{\text{Width of old fabric}}{\text{Width of new fabric}} = \frac{\text{New yarn number}}{\text{Old yarn number}}$$

Stitch density

$$= \frac{\text{Constant}}{\text{Stitch length in inches}}$$

Length constant
= Courses per inch × Stitch length in inches
Width constant
= Wales per inch × Stitch length in inches
Area constant – (Stitch length in inches)^2 × stitch density
= Courses per inch × Wales per inch × (Stitch length in inches)^2
= Length constant × width constant
Loop shape factor

$$= \frac{\text{Length constant}}{\text{Width constant}}$$

$$= \frac{\text{Courses per inch}}{\text{Wales per inch}} = 1.3$$

Weight per sq. inch of fabric in lb – stitch density × stitch length in inches × weight in lb. of 1 inch yarn
Fabric width

$$= \frac{\text{Number of needles}}{\text{Wales per inch}}$$

$$= \frac{\text{Number of needles}}{\text{Width constant/stitch length inches}}$$

(Width constant = Wales per inch × Stitch length in inches)

$$= \frac{\text{Number of needles} \times \text{Stitch length in inches}}{\text{Width constant}}$$

Weight per running yard

$$= \frac{\text{Weight per sq. yard} \times \text{Width of the fabric lb}}{\text{Width constant}}$$

$$= \frac{36 \times \text{Constant} \times \text{Weight in lb of 1 inch of yarn} \times \text{Number of needles}}{\text{Width constant}}$$

= 36 × Length constant × Weight of lb of one inch of yarn in lbs × Number of needles.

Weight per square yard in rib structure (in ozs)

= 2.06 × Courses per inch × Wales per inch × Stitch length in inches/Worsted count

Cover factor or tightness factor

$$= \frac{1}{\text{Stitch length in inches} \times \text{worsted count}}$$

(indirect system)

Or $\dfrac{\text{Tex}}{\text{Stitch length in inches}}$ (direct system)

20.7   Calculations pertaining to warp knitting

Unlike weft knitting, the unit of production in case of warp knitting is a rack.

1 rack = 480 courses of knitting

The length of yarn consumed in knitting one rack is called a 'runner' or a 'run in'

Pounds per rack

= Machine gauge × No. of needle bars × total warp ends

$$= \frac{\text{Width (inches)} \times \text{Average in run-in length} \times \text{Yarn denier}}{36 \times 840 \times 5315}$$

Pounds per linear yard

$$= \frac{\text{Courses per rack}}{\text{Inches per rack}}$$

Courses per inch

$$= \frac{\text{No. of courses/rack} \times \text{Unraveled length of back bar yarn (inches)}}{\text{No. of courses} \times \text{Weight of sample (grams)}}$$

Pounds per hour

= Racks per hour × Pound per rack

Length of yarn in one rack

$$= \frac{\text{Weight of front bar rack} \times \text{length of yarn in one rack}}{\text{Weight of length of yarn in one inch space of given denier}}$$

Weight of front bar yarn

= Weight/sq. inch of sample cloth − Weight of length of yarn in one inch space of given denier

Length of front bar per rack

$$= \frac{\text{Weight of front bar rack} \times \text{length of yarn in one rack}}{\text{Weight of length of yarn in one inch space of given denier}}$$

Number of racks per hour

$$= \frac{\text{RPM} \times 60}{\text{No. of courses/rack}}$$

Production in lbs/hour (inches)

= No. of racks × yarn denier × runner length

$$= \frac{\text{No. of ends} \times \text{efficiency \%}}{9846 \times 36 \times 453 \times 100}$$

21.1 Introduction

Fabric testing includes the testing of various parameters of the fabric. These may be related to the fabric dimensions or properties. The dimensions to be measured are length, width and thickness. The fabric properties include resistance to abrasion, shrinkage, strength, drape, etc.

In this chapter various formulae relating to fabric testing are mentioned. A number of definitions are given to understand the concept of the formulae used.

21.2 Useful definitions

Thread density

It indicates the number of threads of either warp or weft or both, in a given unit area of the fabric.

Crimp

It is the waviness of the yarns in a fabric caused to the interlacement with each other and is normally expressed as a percentage.

Cover

This is a measure of the relative closeness of the yarns in a fabric. Higher the thread spacing, lower the cover of the fabric and vice versa.

Areal density

It is the mass per unit area of the fabric and is expressed in terms of grams per square meter.

Drape

It is the ability of the fabric to assume a graceful appearance when in use.

Bending length

It is the length of fabric that will bend under its own weight to a definite extent; it is a measure of the stiffness that determines draping quality.

Flexural rigidity

It is a measure of stiffness associated with handle.

Bending modulus

It is a value independent of the dimensions of the fabric strip tested and may be regarded as the intrinsic stiffness.

21.3   Miscellaneous formulae

$$\text{Fabric assistance} = \frac{\text{Strip strength per thread}}{\text{Single thread strength}}$$

In a ballistic test,

$$\text{Tearing strength} = \frac{1}{6}(W4 - W1)\,\text{lbs}$$

Let, W_1 = mean tearing energy of five specimens torn through 1 inch and W_4 = mean tearing energy of five specimens torn through 4 inches.

In an Elmdorf tearing tester,
Tearing strength in grams

$$= \text{Capacity of instrument} \times \frac{\text{Pointer reading}}{100}$$

In a BFT abrasion tester,

$$\text{Flex result} = \frac{\text{Mean value of 5 tests}}{1000}$$

$$\text{Ball toughness} = \frac{\text{Mean of 5 tests}}{1000}$$

$$\text{Flat resistance} = \frac{\text{Mean of 5 tests}}{1000}$$

The duty factor is the harmonic mean of ball and flex abrasion results, and is given by

$$\text{Duty factor} = \cfrac{2}{\cfrac{1}{\text{Ball toughness}} + \cfrac{1}{\text{flex result}}}$$

The Figure of merit is the harmonic mean of the ball, flex and flat abrasion.

$$\text{Figure of merit} = \cfrac{3}{\cfrac{1}{\text{Ball toughness}} + \cfrac{1}{\text{flex result}}}$$

The drape coefficient is given by,

$$F = \frac{As - Ad}{A_D - Ad}$$

where A_D = area of the specimen
Ad = area of the supporting disc
As = actual projected area of the specimen

or

$$F = \frac{W_s - W_d}{W_D - W_d}$$

where W_D = Weight of the paper whose area is equal to the area of the specimen.
Wd = Weight of the paper whose area is equal to the area of the supporting disc.
W_s = Weight of the paper whose area is equal to the projected area of the specimen.

In a fabric stiffness tester,

$$\text{Bending length of a fabric, } C = lf(\theta)$$

where l = length of overhang when it is depressed under its own weight
θ = the angle between the line joining the tip to the edge of the platform

$$[f(\theta) = (\cos] \quad [(1/2] \quad \theta / 8)^{1/3}$$

Flexural rigidity of a fabric = 3.39 w_1 c^3 mg.cm
$w^2 c^3 \times 10^3$ mg.cm

where c = bending length in cm
w_1 = weight/sq. td of fabric in ounces
w_2 = weight/sq. cm of fabric in grams

Overall flexural rigidity

$$= (\text{warp way flexural rigidity} \times \text{weft way flexural rigidity})^{1/2}$$

Bending modulus of fabric

$$= 732 \times \text{flexural rigidity kg/sq cm (fabric thickness in thousands of an inch)}^3$$

$$= \frac{12 \times \text{flexural rigidity} \times 10^{-6} \text{ kg/sq cm}}{(\text{fabric thickness in cm})^3}$$

bending length of a limp fabric $= I_0 f(\theta)$

Where $q = 32.85 \times d/I_0$ degrees
$D = I - I_0$
I = Actual length of the loop
Io = 0.1337 L
L = Specimen length in cm
$f(\theta) = (\cos\theta / \tan\theta)^{1/3}$

$$\text{Air permeability} = \frac{\text{Average rate of air flow}}{5.07} \text{cc/s/sq cm}$$

$$\text{Air resistance} = \frac{5.07}{\text{average rate of air flow}} \text{ seconds}$$

21.4   Calculations for cloth

21.4.1 Introduction

This chapter deals with fabric calculations. The fabric calculations comprise of warp and weft calculations. Any fabric that is designed for a particular end use should possess certain properties, such as strength, flexibility, porosity, wear resistance, fineness, and many others. These properties are determined to a great extent by the structure of the fabric, which is characterized by the fabric setting, the yarn properties, the weave, the fabric cover, the yarn crimp percentage, and mass per unit area and so on.

21.4.2 Important parameters

(a) Fabric setting

The term sett is used to indicate the spacing of threads in cloth. Usually the set is the number of threads per 1 cm, or per 10 cm. It is determined for grey as well as for finished cloth. Any cloth is characterized by the number of warp threads (or ends) and the number of weft threads (picks per cm) per cm. The terms "warp density" and "weft density" are also used. The density can also be expressed by the distance between the axis of the adjacent threads, i.e. the spacing, s, which is the reciprocal of the sett and can be found by the formula

$$s = 100/P$$

Where P is the number of threads per 100 mm, and 's' is the distance from center to center of the threads in mm.

Cloth sett is usually given in pairs, warp and weft, warp × weft, $P_1 \times P_2$, where P_1 is the warp density, and P_2 is the weft density. In square sett or balanced fabrics, $P_1 = P_2$.

In designing a new fabric, it is common to calculate the maximum possible density which can he achieved for the given warp and weft yarns.

The theoretical maximum density is obtained when there is no space between the adjacent threads. In reality the density of the threads, at least in one of the systems, can he greater than the theoretical maximum due to compression and bending of threads.

(b) Yarn linear density

For producing fabrics, thinner or thicker yarns can be used, depending on the end use. In the tex system the thickness of yarn is specified by the linear density. The linear density of yarn in tex is the mass in grams of one kilometer of yarn. Linear densities of yarn in the fabric are often given in pairs, warp × weft, $T_1 \times T_2$, where T_1 is the warp linear density, and T_2 is the weft linear density.

In calculations, the diameter of yarn is often used in this case, the yarn is considered as a cylinder with a circular cross section of diameter 'd'. If the length of the cylinder is 'l', its volume is calculated as $v = \pi d^2 l/4$. To find the mass of yarn, m, the second assumption should be done concerning the density of yarn, ρ. The density of yarn is determined by the density of the fibre and the density of the fibre and the percentage of air in the yarn, which depends on a number of factors and is assumed usually as 40–45%. For cotton yarn of low or medium twist the density is assumed to be 0.8 g/cm³, and for high twist 0.9 g/cm³. The diameter of yarn can be calculated as follows, in mm

$$d = \frac{\sqrt{4T}}{\sqrt{\pi\rho 1000}}$$

The factor of 1000 is introduced because tex is defined in terms of g/km.

When the yarn twist is low, $\rho = 0.8$ g/cm^3, and the linear density, $T = 1$ tex. The yarn diameter in mm in the case of cotton is given as,

$$d = 0.0357 \ \sqrt{T/\rho}$$

This formula can be used for yarns with specific volume 1. 1 g/cm^3/g. 11w Specific volume is a reciprocal of yarn density and is determined as $1/\rho = 1/0.9 \ 1 = 1.1$ cm^3/g. When either the fibre density or air content, in the yarn is changed, the formula should he refined.

Sometimes, the thickness of yarn is specified as a metric count, which is the yarn length in meters that weighs in gram. In this case, the formula Nm for calculating the yarn diameter in mm is

$$d = 1.18 - Nm$$

According to the English system, the yarn diameter in inches is calculated by the formula

$$d \ (in.) = 1/28 - Ne$$

(c) Crimp of yarn in a woven fabric

Fabric is produced by interlacing of warp and weft threads. Interlacing causes the bending of the threads round each other. Due to this the warp and weft threads have a wavy shape in the fabric. The wavy shape of the threads can be estimated either by crimp or by take-up.

The crimp is calculated by expressing the difference between the straightened thread length and the sample length, as a percentage of sample length

$$\text{Crimp \%} = \frac{\text{Straightened length} - \text{Sample length}}{\text{Sample length}} \times 100$$

or
$$\text{Crimp\%} = \frac{\text{Uncrimped length} - \text{Crimped length}}{\text{Crimped length}} \times 100$$

Crimp shows the excess of thread length because of curvature of the thread. Take up is calculated by expressing the difference between the straightened thread length and the sample length as a percentage of straight or non-woven length of yarn, and it shows the loss in the length of the thread in weaving.

$$\text{Take-up \%} = \frac{\text{Tape length} - \text{Cloth length}}{\text{Tape length}} \times 100$$

Usually the take up is determined in the preparatory department of a weaving factory for calculating the greater length of yarn necessary for producing a certain length of fabric, i.e. the amount of warp and weft to be ordered for making a fabric of a given length.

Crimp of warp and weft threads should be measured in woven fabric. Crimp of yarn in a particular fabric depends on the sett and the yarn linear density. The ratio of warp and weft crimps is important, because of its great influence on the fabric properties, such as strength, elongation, and others, and on the ratio of a certain property in warp and weft direction.

There is a close relation between the ratio of crimps and the thickness of fabric. By changing this ratio, the mutual displacement of the warp and weft threads, normal to the plane of the fabric, takes place.

(d) Fabric cover

One of the main characteristics of fabric is the density of yarns or yarn spacing. But in some cases, such as, in filter fabrics, for example, this characteristic is not sufficient, because the space between the adjacent threads depends on the yarn thickness. Due to this, the yarn diameter should be taken into consideration. It is known that the fabric with the same density of threads may have different spaces between the threads because of the difference in diameters. On the contrary, the fabric with different densities of threads may have the same space between the threads when the smaller density is combined with a greater diameter. Therefore, the relative closeness of threads depends on the density of threads and their diameters.

The cover reaches the maximum value when the threads cover the whole fabric area. The fabric cover is the ratio of fabric area covered by warp and weft threads to the total fabric area. The maximum theoretical value of cover factor is 28.

(e) Fabric areal density

The mass per unit area, which is termed "areal density", is an important characteristic of the fabric. The mass in the international system of units can be expressed in grams per square metre. To calculate the areal density of fabric, the following factors should be given: the warp and weft linear densities, setts, and crimps. The areal density is calculated by finding the sum of the mass of warp per square metre and that of weft per square metre. The mass of the threads can be found taking into consideration the crimp or take up. When the crimp percentage is used, the formula for calculating the mass of warp in grams per square metre is

Mass of warp in grams

$$= \text{Warp linear density (tex)} \times \text{Warp sett } (1 + 0.01 \times \text{warp crimp\%}) \times 10^{-2}$$

The factor of 10^{-2} is introduced because tex is defined in terms of g/km, and, therefore, the length in m should be converted into km. And the warp sett is multiplied by 10 to convert dm into m. So, the resultant factor is 10^{-2}.

The mass of weft per square metre is given by

Mass of weft in grams

$$= \text{Weft linear density (tex)} \times \text{Weft sett } (1 + 0.01 \times \text{weft crimp\%}) \times 10^{-2}$$

The areal density of fabric in g/m^2 is the sum of the masses of both the warp and weft in grams.

There is a certain relationship between the areal density and the cover factor. The cover factor in the tex system is

$$\text{Cover factor} = \text{Cloth sett } \sqrt{\text{Linear density in tex}}$$

21.5   Other miscellaneous formulae

Cut length or tape length

$$= \text{Required length of cloth} + \text{uptake}$$

$$= \frac{\text{Required length of cloth} \times 100\% + \% \text{ of regain of warp}}{100}$$

Reed width

$$= \frac{\text{Cloth width} \times 100 + \% \text{ of regain}}{100}$$

Reed count

$$= \text{Number of ends/inch in the cloth} \times \frac{\text{Width of cloth}}{\text{Reed width}}$$

Loom pick

$$= \text{Pick per inch in cloth} \times \frac{\text{Length of cloth}}{\text{Tape length}}$$

21.6   Warp calculations

Total number of ends in cloth

$$= \text{Number of ends per inch in cloth} \times \text{Cloth width} + \text{Extra ends for selvedge}$$

Or   number of ends per inch in reed × Width of warp in reed +
          Extra ends for selvedges

Length of warp yarn in the cloth in yards

$$= \text{tape length in yards} \times \text{number of ends}$$

Total length of warp yarn in hanks

$$= \frac{\text{Total number of ends} \times \text{Tape length in yards}}{840}$$

Weight of warp in lbs

$$= \frac{\text{Total length of warp yarns in hanks}}{\text{Count}}$$

$$= \frac{\text{Total number of ends} \times \text{Tape length of warp in yards}}{840 \times \text{Count}}$$

Count of warp yarn

$$= \frac{\text{Total length of warp yarn in hanks}}{\text{Weight of warp in lbs}}$$

$$= \frac{\text{Total number of ends} \times \text{Tape length in yards}}{840 \times \text{Weight of warp in lbs}}$$

Tape length in yards

$$= \frac{\text{Count of warp yarn} \times \text{Weight of warp in lbs} \times 840}{\text{Total number of ends}}$$

Length of cloth in yards

$$= \frac{\text{Count of warp yarn} \times \text{Weight of warp in lbs} \times 840 \text{ uptake}}{\text{Total number of ends}}$$

(Cloth length = Tape length – Uptake)

Total number of ends

$$= \frac{\text{Counts of warp yarn} \times \text{Weight of warp in lbs}}{\text{Tape length in yards}} \times 840$$

Weight of warp in lbs

$$= \frac{\text{Tape length in yards} \times \text{Total number of ends in body warp} + \text{Tape length in yards} \times \text{Total number of ends in selvedges}}{840 \times \text{Count of body warp} \times 840 \times \text{Count of selvedge}}$$

$$= \frac{\text{Tape length in yards (total number of ends of body warp} + \text{Total number of ends for selvedges)}}{840 \times \text{Count of body warp} \times \text{Count of selvedge yarn}}$$

When warp contains yarns of different counts

Weight of warp in lbs

$$= \frac{\text{Tape length in yards (total no. of ends of yarn A + total number of ends yarn B)}}{840 \times \text{Count of yarn A} \times \text{Count of yarn B}}$$

21.7   Weft calculations

Width of warp in reed1n inches

$$= \frac{\text{Cloth width inches} \times \% \text{ of regain of weft} + 100}{100}$$

Total length of weft in yards
= Total number of picks × Length of weft in reed in yards
= Length of cloth in yards × Picks per inch in cloth ×
                                    Reed width in inches

Total length of weft in hanks

$$= \frac{\text{Length of cloth in yards} \times \text{Picks per inch in the cloth} \times \text{Reed width in inches}}{840}$$

Weight of weft in lbs

$$= \frac{\text{Total length of weft yarns in hanks}}{\text{Count of weft}}$$

$$= \frac{\text{Length of cloth in yards} \times \text{Picks per inch in the cloth} \times \text{Reed width in inches}}{840 \times \text{Weight of weft in lbs}}$$

Length of cloth in yards

$$= \frac{\text{Count of weft yarn} \times \text{Weight of weft in lbs}}{\text{Picks/inch in cloth} \times \text{Reed width in inches} \times 840}$$

Picks/inch in cloth

$$= \frac{\text{Count of weft in lb}}{\text{Cloth length in yards} \times \text{Reed width in inches}} \times 840$$

Reed width in inches

$$= \frac{\text{Count of weft} \times \text{Weight of weft in lbs}}{\text{Cloth length in yards} \times \text{Picks/inch in cloth}} \times 840$$

When weft yarns of different counts are used

Weight of weft in lbs

$$= \frac{\text{Length of weft } A \text{ in hanks} + \text{Length of weft } B \text{ in hanks}}{\text{Count of weft yarn } A \times \text{Count of weft yarn } B}$$

$$= \frac{\text{Length of cloth in yards} \times \text{Reed width in inches} \times \text{Picks/inch in cloth}}{840 \times \text{Average count of weft}}$$

21.8   Quantity of material in a piece

Total weight of yarn in a piece

= Weight of warp yarn + Weight of weft yarn

Total weight of a piece

= Weight of warp yarn + Weight of weft yarn + Weight of size or finishing materials

21.9   Calculations related to cloth sett

The maximum setting for a square plain cloth

$$= \frac{\text{Number of threads/repeat} \times \text{Reciprocal of diameter of yarn}}{\text{Number of intersections/repeat} + \text{Number of threads/repeat}}$$

For a plain cloth the maximum setting

$$= \frac{2 \times \text{Reciprocal of diameter}}{2 + 2}$$

$$= \frac{2 \times \text{Reciprocal of diameter}}{4}$$

$$= \frac{\text{Reciprocal of diameter}}{2}$$

Ends/inch required in cloth

$$= \frac{\text{Ends/inch in given cloth} \times \% \text{ count of yarn in required cloth}}{\text{Count of yarn in given cloth}}$$

(Ends/inch in the required cloth)^2

$$= \frac{(\text{Ends/inch in given cloth})^2 \times \text{Count of yarn in required cloth}}{\text{Count of yarn in given cloth}}$$

Ends/inch in given cloth $\times \sqrt{\text{count of yarn in required cloth}}$

$= $ Ends/inch in required cloth $\times \sqrt{\text{Count of yarn in given cloth}}$

Therefore, count of yarn in required cloth

$= $ (Ends/inch in required cloth/ends/inch in given cloth)$^2 \times$
Count of yarn in given cloth

Number of diameters/inch in required cloth

$= $ Number of diameters/inch in known cloth

or Number of threads/inch in required cloth + Number of intersections/
inch in required cloth

$= $ Number of threads/inch in known cloth + Number of
intersections in known cloth

Number of intersections/inch in a cloth

$$= \frac{\text{Number of threads/inch } (\div) \text{ Number of threads/repeat}}{\text{Number of intersections/repeat}}$$

$$= \frac{\text{Number of threads/inch} \times \text{Number of intersections per repeat}}{\text{Number of threads/repeat}}$$

Therefore, threads/inch in required cloth + number of threads/inch in

$$= \text{Required cloth} \times \frac{\begin{array}{c}\text{Number of intersections/ repeat in}\\ \text{required cloth}\end{array}}{\begin{array}{c}\text{Number of threads in required}\\ \text{cloth}\end{array}}$$

Number of threads/inch in required cloth + Number of threads/inch

$$= \frac{\begin{array}{c}\text{in known cloth} \times \text{Number of intersections/repeat in}\\ \text{known cloth}\end{array}}{\text{Number of threads in known cloth}}$$

Number of threads/inch in required cloth (1 + Number on intersections/
repeat in required cloth/number of threads/repeat in required cloth)

= Number of threads/inch in known cloth (1 + Number of intersections per repeat in known cloth/number of threads/ repeat in known cloth)

(or) $\dfrac{\text{Number of threads/inch in required cloth}}{\text{Number of threads/inch in known cloth} + 1}$

$$= \frac{\begin{array}{c}\text{Number of intersections/}\\ \text{repeat in known cloth}\end{array}}{\begin{array}{c}\text{number of threads/repeat}\\ \text{in required cloth}\end{array}} + \frac{\text{Repeat in known cloth}}{\begin{array}{c}\text{number of intersections/}\\ \text{repeat in required cloth}\end{array}}$$

Number of threads/inch in known cloth × Number of threads/ repeat in required cloth ×

$$= \frac{\begin{array}{c}\text{Number of threads/}\\ \text{repeat in known cloth}\end{array}}{\begin{array}{c}\text{(Number of threads/}\\ \text{repeat in known cloth)}\end{array}} + \frac{\begin{array}{c}\text{Number of intersections}\\ \text{in known cloth}\end{array}}{\begin{array}{c}\text{(Number of threads/}\\ \text{repeat in required cloth +}\\ \text{Number of intersections in}\\ \text{required cloth)}\end{array}}$$

Cloth cover factor, $K = \dfrac{k_1 + k_2 - k_1 k_2}{28}$

Where K = Cloth cover factor
k_1 = Warp cover factor
k_2 = Weft cover factor

$$k_1 = \frac{n_1}{\sqrt{N_1}} \; ; k_2 = \frac{n_2}{\sqrt{N_2}} \text{ (indirect system)}$$

$$k_1 = n_1\sqrt{N_1} \; ; k_2 = n_2\sqrt{N_2} \text{ (direct system)}$$

where n_1 = ends/inch
n_2 = picks/inch
N_1 = warp yarn number
N_2 = weft yarn number

Statistics for quality control

22.1 Basic statistics

Statistics is mainly applicable to testing in spinning besides other processes such as winding and weaving. It is necessary in process to process analysis for variation and to keep the product quality under acceptable limits and to keep the disturbing influences in the process under control to achieve maximum production. Knowledge of statistics is particularly essential for every spinner.

22.2 Key definitions

Population

It is the total set of items defined by some characteristics of those items. For example, Doff weight of 60^s yarn in a ring frame.

Sample

It refers to any finite set of items drawn from the population. For example, 5 cops drawn from the Doff of a frame.

Random sample

It is a sample from given population each clement of which has got equal chance of being drawn from the population.

Frequency distribution

It is constructed from minimum of 40 readings to 100 and 200 items. Mode and media are not commonly used in textile industry.

Attribute

Any selected quality or characteristic (that is present or absent in each of the units) that has to be statistically treated and judged is known as an attribute.

Variate

It is the particular quality whose actual magnitude of character is measured from different units; for example, the weight of a group of units. The quantity which varies (height, weight, income, etc.) is called the variate.

Variability

It is the variation of test results about a central value. Amongst the variables there are continuous as well as discontinuous variables.

Dispersion

It is almost the same as variability.

Scatter

Like dispersion scatter also means variability.

Frequency

It is the number of times a particular characteristic occurs in a given sample or during a set interval, etc., which also implies the number of observations lying between two specified limits.

Frequency distribution

It is the complete range of data arranged in a table of frequencies.

Frequency polygon

It is the frequency distribution of a sample or population represented by a graphical method.

Histogram

It is a grouped frequency distributed when represented by a series of rectangles side by side. This is an alternative method of diagrammatic representation.

Frequency curve

It is a smooth curve obtained from a histogram, when the number of observations becomes very large.

Mean

It is the same as usual arithmetic average where the sum of observations is divided by the number of observations.

Deviation

It statistically implies the difference between an observation and the mean of all observations.

Median

It is the middle value of a set of test values. The no. of test values greater than and lesser than are same. The observations are usually arranged in an increasing order.

Mean deviation

It is obtained when the sum of all the differences of the values from their mean, taken without regard to sign, is divided by the number of values. This mean deviation is a measure of the spread.

Range

It is the difference between the maximum and minimum values in a set of data and is the simplest measure of spread.

Variance

It is the square of all deviation (all signs become +ve) and their mean (i.e. the mean of all squared deviations) or in other words it is the squared standard deviation.

Standard deviation

It is defined as the root mean square deviation of numbers from their average. Square root of variance is also same as standard deviation and this is a useful tool in statistical results. The commonly used notations are Σ (sigma) or δ (delta) or by SD. The SD for a population is the square root of the average of the squares of the deviations by one less than the numbers of observations before taking the square root. This is an essential step in the condensation of data. In other words, the SD is a measure of the amount of variation I from the average of a set of data.

Estimate

It is a value calculated from a sample to appraise a characteristic of the population.

Mode

It is that particular value which occurs most frequently in a set of data.

Coefficient of variation

It is a very useful tool in statistical studies and is usually denoted by CV. It is obtained when standard deviation is expressed as a percentage of the mean. CV is usually employed to compare the relative variations in different sets of data. Thus the CV is a measure of the relative dispersion and is useful in comparing the dispersions of two or more processes or materials or in comparing the same types of materials produced at different times.

Control chart

It is a chart that is prepared to detect abnormalities and displaying the results.

Correlation

It shows the interdependence of two variables.

Limits of variation

It is the range within which all hut a specified small proportion of the values fall.

Tolerance limit

It is the permissible limit of variation.

Percentage mean deviation

It is a deviation expressed as a percentage of the mean and is denoted by PMD.

22.3   Importance of SD

The SD is an ideal tool, particularly for a spinning mill to measure the degree of variations in raw material, yarn, men and machinery. The use of SD is illustrated in the example given below:

Difference from average	Square of difference	Frequency	Total
4	16	1	16
3	9	2	18
2	4	3	12
1	1	4	4
0	0	5	0
1	1	4	4
2	4	3	12
3	9	2	18
4	16	1	16
		25	100

$$\text{Standard deviation} = \frac{\sqrt{100}}{25} = 2$$

After obtaining SD, the coefficient of variation can be obtained by the formula:

$$CV\% = \frac{SD}{\text{Average}} \times 100 = \frac{2}{43} \times 100 = 4.65\%$$

22.4   Practical importance of CV%

(a) Ring frame count and strength wrappings
(b) Fabric strength
(c) Cone winding count wrappings
(d) Draw frame wrappings

22.5   Average range method for calculating SD

This method avoids laborious calculations and hence mills adopt this method to calculate the SD directly.

For example, 10 finisher draw sliver wrappings were taken and the following results were obtained:

(a) Total of 10 individual readings is 560 grains/yard.
(b) Average of these 10 readings or grand average is 56 grains/yard.
(c) The total of all ranges is 32.5 grams
(d) The average range = 10 = 3.25

While wrapping was taken, a group of 5, i.e. 5 readings formed one group and for each 5 readings, the average and range were worked.

The table below is used for easily obtaining SD directly for each group. For a group of 5 the factor is 0.43.

Multiply the factor with average range = 3.25 × 0.43 = 1.3975

$$\text{Coefficient of variation} = \frac{1.3975}{56} \times 100 = 2.496$$

22.6   Variation analysis

This is used in spinning mills, where two types of variation are encountered, namely

(a)   within machine variation, and (b) between machine variation.

Sample consists of the following number of units	Obtain SD by multiplying average range by
2	0.89
3	0.59
4	0.49
5	0.43
6	0.40
7	0.37
8	0.35
10	0.33
12	0.31
14	0.29
16	0.28

In addition to the above, overall variation gives the section variation of the particular group.

22.6.1 Within delivery variation

This type of variation occurs when successive strands of wrappings from single strand or length like yard to yard from scutcher lap is taken. But this is reduced through number of doubling.

22.6.2 Within machine variation

In drawing when wrapping from single delivery is taken, this type of variation occurs. To get representative sample, several deliveries from the frame for the count group have to be taken.

22.6.3 Between machine variations

This can be found by computing CV% from among frame averages.

22.6.4 Overall variation

Samples are taken from several machines running on the same count.

22.7   Process to process variation analysis

This type of analysis is useful to find out whether the variation introduced in succeeding process is normal and well within limits.

A mathematical formula is available to predict the expected variation.

Let V_0 and V_m be the overall and within machine variation respectively and N the number of doubling introduced, then,

$$V_m = \frac{V_0 \text{ of the previous process}}{\sqrt{N}}$$

For example, the variation of card sliver is 5%. With 8 of doubling it was passed through draw frame. Then the expected variation in the draw sliver will be

$$\text{Expected draw sliver variation} = \frac{5}{\sqrt{8}} = 1.76$$

Continuing the same with 8 of doubling in the II passage of draw frame:

$$\text{II passage sliver variation} = \frac{1.765}{\sqrt{8}} = 0.622$$

For calculation of the process variation, samples have to be selected at random, and should be representative. In order to find out whether the results arc significant, the following procedure has to be applied:

Standard error = where N is the number of individuals in the sample

Example: A mill tested 50 cops for count and strength. The CV of count is 5% and the standard deviation is 2.6, then the standard error will be x. The average count tested from 50 leas would vary from 40 ± 0.66 in 95% of the cases (Nominal count = 40s)

$$\text{Standard error} = \frac{2.6}{\sqrt{50}} = 0.367$$

22.8   Sample size

In order to correctly assess whether the result is reliable, sample size should be correctly determined for collecting the data.

$$\text{Sample size, } N = \frac{4 \times CV\%}{e^2}$$

where, e – error allowed.

In a spinning mill, testing is done daily, deficiencies observed and corrective actions are taken to improve the quality and reduce the end breakages. In blow room, improve the CV% of lap, improve the card sliver quality by reducing the neps, improve the U% in drawing by resetting, etc. Readings are taken before and after. Test of significance is a tool which provides a basis to conclude that there is improvement.

Test of Significance – Average: 'U' test and 'T' test

To determine whether the average determined from a sample is significantly different from the population average: If x is the sample average, p the population average and v^2 is the population variance, then,

$$U = \frac{x - p}{v / \sqrt{n}} \text{ , where } n \text{ is the sample size}$$

Example: 60s yarn was tested for lea strength and the average of 25 leas gave 12 kg, and standard deviation of 2, whereas the specification is 13 kg. It is possible to accept that the yarn has an average of 13 kg arid pass on?

$$U = \frac{12 - 13}{2 / \sqrt{25}} = \frac{1 \times 5}{2}$$

This exceeds 1.96 hence it is significant and the yarn cannot be passed. When population variation is not known, then "T" test is applied.

Formula:

1. $t = \dfrac{\overline{x} - \overline{x}}{v / \sqrt{n}}$ and

2. $t = \dfrac{\overline{x_1} - \overline{x_2}}{\dfrac{v / 1}{\sqrt{n_1}} + \dfrac{1}{\sqrt{n_2}}}$

The value of t calculated and if exceeds the value of the 't' table at 95% confidence limits, then there is significant difference.

22.9   Test of significant variance

If standard deviations of 2 samples arc known, we can find whether the samples have significant difference of variation by applying 'F' test.

Formula: $F = \dfrac{S_1^2}{S_2^2}$

The result obtained is compared with 'F table for $(n_1 - 1)$ and $(n_2 - 1)$ table at 95% confidence limit. If the value lies below the reading then it is not significant.

22.10   Control chart

The control chart is a useful statistical tool to regulate changes in processes. Important areas of application are wheel changes in drawing, twist variation, end breakages and waste extraction in carding.

22.11   Methods of computing 'average' and 'range' charts

1. Collect data in sub-groups of size n. Build up the data to 25 to 30 subgroups say k. Total of 100 to 200 readings spread over different days will be ideal.

2. Calculate the average and range of each sub group.

3. Find the grand average: $\bar{x} = \dfrac{X_1 + X_2 + ...X_k}{k}$

4. Calculate the range: $\bar{R} = \dfrac{R_1 + ... R_k}{k}$

5. Calculate the upper and lower limits for the sub-groups by referring the statistical table $D4\bar{R}$ and $D3\bar{R}$.

6. If any group range is larger than $D4\bar{R}$ and $D3\bar{R}$ such readings are omitted and the procedure repeated.

7. The final $\bar{R}$ obtained is used to construct the average chart. R/d_2 is the standard deviation of the population.

$$\frac{\bar{x} + 3R}{d_2 / n} = \bar{x} + A_2R$$

The factor d_2 and A_2 can be obtained from the table.

Example: The final head draw frame wrapping gives an average of 60 grains/yard and the range was 4. The sub-group is 3.

$$\text{Average: } X = X + A_2 R = UCL$$

$\overline{X} = X - A_2 R = LCL$. From the above table the factor for group of 3 is 2.58 and for the average is 1.02.

Sample size (n)	Factors for average A_2		Factors for Range D_4	
2	1.88		0	3.27
3	1.02		0	2.58
4	0.73		0	2.28
5	0.58		0	2.11
6	0.48		0.00	2.00
7	0.42		0.08	1.92
8	0.37		0.14	1.86
9	0.34		0.18	1.82
10	0.31		0.22	1.78
11	0.29		0.26	1.74
12	0.27		0.28	1.72
13	0.25		0.31	1.69
14	0.24		0.33	1.67
15	0.22		0.35	1.65
16	0.21		0.36	1.64
17	0.20		0.38	1.62
18	0.19		0.39	1.61
19	0.19		0.40	1.60
20	0.18		0.41	1.59
Sub-group	**X-chart**		**LCD**	**UCD**

Average = 60 + (4 × 1.02) UCL = 64.08
Range = 4 × 2.58 = 10.32 grains UCL.

Chart for 't' test and '÷²' test

Degree of freedom	't' Distribution		χ^2 Distribution	
	5%	1%	5%	1%
1	12.71	63.66	3.84	6.63
2	4.30	9.92	5.99	9.21
3	3.18	5.84	7.81	11.34
4	2.78	4.80	9.49	13.28
5	2.57	4.03	11.07	15.09
6	2.45	3.71	12.59	16.81
7	2.36	3.50	14.07	18.48
8	2.31	3.25	16.02	21.67
10	2.23	3.17	18.31	23.21
11	2.30	3.11	19.68	24.72
12	2.18	3.06	21.03	26.22
13	2.16	3.01	22.36	27.69
14	2.14	2.98	23.68	29.14

(Contd.)

Degree of freedom	't' Distribution		χ^2 Distribution	
	5%	1%	5%	1%
15	2.13	2.95	25.00	30.58
16	2.12	2.92	26.30	32.00
17	2.11	2.90	27.59	33.41
18	2.10	2.88	28.87	34.81
19	2.09	2.86	30.14	36.19
20	2.09	2.84	31.41	37.57
21	2.08	2.83	32.67	38.93
22	2.07	2.82	33.92	40.29
23	2.07	2.81	35.17	41.64
24	2.06	2.80	36.42	42.98
25	2.06	2.79	37.65	44.31
26	2.06	2.78	38.89	45.64
27	2.05	2.77	40.11	46.96
28	2.05	2.76	41.34	48.28
29	2.04	2.76	42.56	49.59
30	2.04	2.75	43.77	50.89
40	2.02	2.70	55.76	63.69
60	2.00	2.66	79.08	88.38

'F' Distribution – Upper 5%

N_2	10	12	15	20	24	30	40	60
1	241.9	243.9	247.9	248.0	249.1	250.1	251.1	252.2
2	19.4	19.4	19.43	19.45	19.4	19.45	19.47	19.48
3	8.79	8.74	8.70	8.66	8.64	8.62	8.59	8.57
4	5.96	5.91	5.86	5.80	5.77	5.75	5.72	5.69
5	4.74	4.68	4.62	4.56	4.53	4.50	4.46	4.43
6	4.06	4.00	3.94	3.87	3.84	3.81	3.77	3.74
7	3.64	3.57	3.51	3.44	3.41	3.38	3.34	3.30
8	3.35	3.28	3.22	3.15	3.12	3.08	3.04	3.01
9	3.14	3.07	3.01	2.94	2.90	2.86	2.83	2.79
10	2.98	2.91	2.85	2.77	2.74	2.70	2.66	2.62
11	2.85	2.79	2.72	2.65	2.61	2.57	2.53	2.49
12	2.75	2.69	2.62	2.54	2.51	2.47	2.43	2.38
13	2.67	2.60	2.53	2.46	2.42	2.38	2.34	2.30
14	2.60	2.53	2.46	2.39	2.35	2.31	2.27	2.22
15	2.54	2.48	2.40	2.36	2.28	2.28	2.30	2.15
16	2.49	2.42	2.35	2.28	2.24	2.19	2.15	2.11
20	2.35	2.28	2.20	2.12	2.08	2.04	1.99	1.95
30	2.16	2.09	2.01	1.93	1.89	1.84	1.79	1.74
40	2.08	2.00	1.92	1.84	1.79	1.74	1.69	1.64
60	1.99	1.92	1.84	1.75	1.70	1.65	1.59	1.53

$$F = \frac{S_1^2}{S_2^2}$$

S_1^2 and S_2^2 are independent mean squares estimating common variance S_2 based on n_1 and n_2 degree of freedom.

In textile mills apart from controlling the wrapping through control charts, control charts are used to control the defects produced whether they are within the limits.

In computing such a chart, a simple formula based on 'Poisson's' formula is used. In adopting this formula, the sample size should be large, more than 50 and the percentage defects should not exceed more than 10.

Formula: Standard deviation = $\sqrt{}$Average

Example: Out of 90 cones checked, on an average 8 cones were found defective. Construct the control chart for maximum allowable defects.

Standard deviation

Upper control limit = Average defects ± (3 × SD)
= 8 + 3 × 3 = 17.
Lower control limit will be zero.

Example: 95 laps were produced in a shift. The proportion of heavier lap was 0.4, if the lap weights are under control. Control limits for heavier lap will be

$$= 0.4 \pm \frac{3 \sqrt{0.4 \times 0.4}}{\sqrt{100}} = 0.4 \pm 0.12$$

Interpretation of the result

= The proportion of heavier lap lies between 35% and 65%, then lap weight is under control.

22.12   Snap reading

No. of readings required for snap reading survey at 95% confidence limit:

$$n = \frac{4p \ (100 - p)}{a^2 N}$$

where, n = number of snaps required
p = approximate efficiency
a = accuracy desired in %
N = number of machines

Example: A mill is having 48 machines. It was desired to have the machine utilization of 95%. How many snaps are to be taken so as to get an accuracy of 1%.

$$\text{No. of readings} = \frac{4 \times 95 \times 5}{1^2 \times 53} = 35.85 \text{ snaps}$$

22.13 Miscellaneous formulae

Class interval

$$= \frac{\text{Largest item} - \text{Smallest item}}{\text{Number of classes}}$$

Arithmetic mean or average X

$$= \frac{X_1 + X_2 + X_3 + \ldots + X_n}{n}$$

where X_1, X_2, X_3, etc. are the individual variable values
n – number of variables

$$\bar{X}_1 = \frac{\Sigma \bar{X}}{N} \text{ (sum of variables)}$$

Mode – mean – 3 (mean – median)
Percentage mean range

$$= \frac{\text{Range}}{\text{Mean}} \times 100$$

Quartile deviation

$$= \frac{\text{Upper quartile} - \text{Lower quartile}}{2}$$

Median – Lower quartile
$$= \text{Upper quartile} - \text{Median}$$
Mean deviation

$$= \frac{\text{Sum of the deviations from the mean}}{\text{Total number of observations}}$$

$$= \frac{\Sigma |X - X^1|}{n}$$

Where, X = Observed value
X^1 = Arithmetic mean
N = Number of observations

Percentage mean deviation

$$= \frac{\text{Mean deviation}}{\text{Mean}} \times 100$$

Standard deviation, SD

$$= \sqrt{S(X^1 - X^1)^2/n - 1}$$

co-efficient of variation, CV

$$= \frac{SD}{X^1} \times 100$$

Variance $= \Sigma \dfrac{(X - X)^2}{n - 1}$

Mean deviation

$$= \text{Standard deviation} \times 0.798$$

Standard deviation

$$= \frac{\text{Mean deviation}}{0.798}$$

Standard deviation – Mean deviation $\times$ 1.25

Standard error of mean

$$= \frac{SD}{\sqrt{n}}$$

Standard error of standard deviation

$$= \frac{SD}{\sqrt{(2n)}}$$

Standard error of coefficient of variation

$$= \frac{CV}{\sqrt{(2n)}}$$

Average fraction defectives

$$= \frac{\text{Total number of defectives}}{\text{Total number of items inspected}}$$

Upper control limit, UCL

$$= n \bar{p} + 3\sqrt{np} (1 - p)$$

Lower control limit

$$= np - 3\sqrt{np} (1 - p)$$

Where n = number of samples inspected

   $\bar{p}$ = average fraction defective

   p = fraction defective

Bibliography

1. Weaving conversion of yarn to fabric — P.R.Lord, M.A.Mohammed.
2. Weaving Calculations — R. Sengupta
3. Spinning Calculations — K. Pattabhiraman
4. Textile Mechanics and Calculations — S. Jagannathan
5. Theory of Machines — R.S. Khurmi
6. Textile Mechanics — Volume 2 — K. Slater
7. Knitting Technology — D.B. Ajgaonkar
8. Textile Testing —J.E. Booth
9. Weaving mechanism – Part I – Chakraborthy
10. Theory of machines , R.S.Khurmi.
11. Mechanics of spinning machines, R.S.Rengasamy, NCUTE Publication, 2003.
12. Weaving mechanism Vol.I, N.N.Banerjee.
13. Textile mathematics, Vol.2 and 3, Textile Institute, UK.
14. Useful formulae in textile calculations, N.Gokarneshan, NCUTE Publication, 2004.
15. Practical guide to fabric manufacture and cloth analysis laboratories, N.Gokarneshan, Mahajan Publishers, 2005.
16. www.srmuniv.edu